WHALES

AND DOLPHINS
IN·QUESTION

WHALES
AND DOLPHINS
IN·QUESTION

THE SMITHSONIAN ANSWER BOOK

JAMES G. MEAD ▸ JOY P. GOLD

PHOTOGRAPHS BY FLIP NICKLIN

SMITHSONIAN INSTITUTION PRESS
WASHINGTON AND LONDON

Copy Editor: Robin Whitaker
Production Editor: Ruth G. Thomson
Designers: Amber Frid-Jimenez, Jana Wicklund, and Janice Wheeler

Library of Congress Cataloging-in-Publication Data
Mead, James G.
    Whales and dolphins in question : the Smithsonian answer book / James G. Mead,
  Joy P. Gold; with photographs by Flip Nicklin.
        p.   cm.
  Includes bibliographical references (p.).
  ISBN 1-56098-955-6 (cloth : alk. paper) — ISBN 1-56098-980-7 (paper : alk. paper)
    1. Cetacea—Miscellanea.   I. Gold, Joy P.   II. Title
  QL737.C4 M42   2002
  599.5—dc21   2001034175

British Library Cataloguing-in-Publication Data available

Printed in China, not at government expense
08 07 06 05 04 03 02 1 2 3 4 5

♾ The paper used in this publication meets the minimum requirements of the American National
Standard for Information Sciences—Permanence of Paper for Printed Library Materials ANSI
Z39.48-1984.

Frontispiece: Killer whales, *Orcinus orca.*

CONTENTS

.2.
WHALE EVOLUTION AND DIVERSITY 95

Photo gallery appears following page 97.

PREFACE

A hundred or so years ago some of the whales that washed ashore on beaches or were killed at sea caught the attention of turn-of-the-century naturalists and scientists. Fortunately, some of the skeletons from those whales were sent to the Smithsonian Institution. Today the Smithsonian has the largest cetacean research collection in the world.

A century ago, the world was interested not in the preservation of whales but rather in the economic profit garnered from whale products. Times have changed to a certain degree. In the instant-information age, within a few hours or even minutes we know that a whale has beached. We have become more aware, more educated, and perhaps more compassionate in our attitudes toward whales. Many nations have prohibited commercial whale harvesting, and laws protecting endangered species often apply to whales. The public is intensely interested in whales and concerned for their well-being, as is evidenced by the thousands of calls and letters received by the Smithsonian's National Museum of Natural History. Indeed, there has been an overall rise of interest in marine mammals, their biology, and their behavior. Here, we attempt to answer the questions about whales that have been posed to the Smithsonian's National Museum of Natural History over the years.

Books, articles, television programs, videos, and movies about these marvelous creatures proliferate. In the information age, where computer-driven communication provides access to all manner of reports and discussions, with a multitude of so-called facts, we are not at all surprised that much information one receives about whales is flawed. We have chosen to write this book to answer the questions that have been addressed to the Smithsonian, to dispel myth and misinformation, and to explore the biology and natural history of whales in a critical and objective manner. We have written this book with the goal of critically examining the lives of these fascinating creatures and responding objectively and accurately to the questions raised about them. Their interesting anatomy and unusual behavior will be discussed, but we will attempt to provide facts where there has been conjecture, dispel myths, and summarize what is known of their ancient history, classification, biology, behavior, and relationships with humans. Along the way, we hope to give you some insight into the diverse world of whales that you may not have had before reading this book.

Humpback whale, *Megaptera novaeangliae.*

Whales and Dolphins in Question is organized in question-and-answer format, similar to its predecessors, *Sharks in Question, Snakes in Question, and Bats in Question.* Most of the questions are answered simply at first, then expanded with additional technical details. There may be repetition in some sections, because we are trying to make this the kind of book in which the reader can look up an answer quickly without having to read the entire text sequentially. The first section, titled "Whale and Dolphin Facts," features the evolutionary history of whales, their anatomy, where and how they feed, breathe, swim, socialize, and reproduce. The second section describes in greater detail various whale species: big ones, small ones, those that have been studied for some years, and those about whom very little is known. Because whales have been and continue to be of economic interest, in the third section we discuss the history of whaling, its effects on present-day stocks, organizations and laws created to protect whales, and the relationship of whales with humans.

This book also contains a glossary of unusual terms and a series of appendixes that list the formal classification of whales, careers in cetology, the principal organizations concerned with whales along with their periodicals, and popular whale bibliographies. A list of technical references relating to cetaceans, their history, and our scientific research on them also appears.

Royalties that would normally accrue to the authors from the sale of this work will instead be deposited into a fund in the Department of Vertebrate Zoology, to be used for the increase and diffusion of knowledge about marine mammals.

ACKNOWLEDGMENTS

The Smithsonian Institution, and in particular the National Museum of Natural History, has fostered an interest in marine mammals since the days when its administrators hired Spencer Fullerton Baird in 1850. This has resulted in an almost unbroken line of research scientists who were concerned with whales, among them Baird, Frederick W. True, Leonard Stejneger, William H. Dall, Gerrit S. Miller, A. Remington Kellogg, Clayton E. Ray, and Charles O. Handley. These scholars contributed to the Smithsonian's mission, "the increase and diffusion of knowledge among men," and have left a collection that is unequaled in terms of its breadth, libraries, and research facilities.

We also greatly appreciate the people who have contributed to the marine mammal stranding program. The stranding program has made it possible to assemble quantities of data, bit by bit, and come up with reasonable conclusions about the natural history of many species of whales and dolphins. We owe a particular debt to the Ford Motor Company, which gave us the means, in the form of a 1973 four-wheel-drive truck, to go out into the field and collect data and specimens of cetaceans. We also thank Nixon W. Griffis for his past support.

Everyone who has been involved with cetaceans has contributed in his or her own special way to make this book what it is. For encouragement and help in numerous ways we would like to thank the following people: Allen Baker, Dick Benson, Dave Bohaska, Bob Brownell, Phil Clapham, Bud Fay, Ewan Fordyce, John Heyning, Toshio Kasuya, Ed Mitchell, Ken Norris, Bill Perrin, John Prescott, Dale Rice, Bill Schevill, Bill Walker, and Tadasu Yamada. Robert Kenney and two anonymous reviewers of the manuscript provided suggestions that improved the work greatly.

Last, we owe extreme thanks to Rebecca Mead, who has put up with one of the coauthors for longer than we care to think about, not only domestically, but in the field as well; to Charley Potter, without whom the Marine Mammal Program would have folded long ago; and to Archie Gold, who, when called upon, patiently gave the other coauthor support in this endeavor and provided technical expertise.

All of the references to the origins of words (etymology) are taken from the *Century Dictionary and Cyclopedia*.

INTRODUCTION

No group of animals has captured the public's mind the way whales and dolphins have. From a little-understood group, seen alive mostly by people working or traveling at sea, they have grown to be a group whose popularity is equaled only by birds.

Research has been done largely on stranded animals. In 1951, the military became interested in dolphin echolocation, and, as a consequence of this interest, funding was accelerated for scientific research. This led to increased facilities for the maintenance of captive dolphins, both in private facilities and public oceanaria. The public became so enthusiastic and attended in such large numbers that boat captains began to offer whale- and dolphin-watching cruises, which are very popular today. Increased sophistication of underwater photographic equipment has given filmmakers the ability to make many fine films of cetaceans at sea, which are now viewed in theaters and on television.

Whales, dolphins, and porpoises inhabit all the world's oceans and many major river systems, from the Arctic to equatorial waters. They range in size from mammoth blue whales at 30 meters and 200 metric tons down to the newborn franciscana (La Plata river dolphin) of 75 to 80 centimeters and 7 kilograms and the finless porpoise of 80 centimeters and 10 kilograms.

Cetaceans have evolved some unique adaptations. The enormous head of the sperm whale is fully a third of the body in length and estimated to be more than a third of the body in weight. It is composed of an extremely modified nose and upper lip! The two teeth of adult male strap-toothed whales grow upward and cross over each other above the upper jaw, permitting the mouth to only open 11 to 13 centimeters at the tip, yet this is an animal that reaches more that 6 meters in length. The narwhal grows a tusk up to 3 meters in length and straight out in front of its head; this tusk can be fully 60 percent of the total length of the body.

In the past, whales and dolphins formed the resources for fisheries and were caught and consumed by man. Now with growing appreciation of the natural environment, whales and dolphins are protected by U.S. law and can be watched, studied, and appreciated for the splendid creatures they are. This philosophy is becoming established in more countries, but some still see whales and dolphins as harvestable resources.

.1.

WHALE AND DOLPHIN FACTS

WHAT ARE WHALES?

Whales are large mammals that live in the marine environment. Like other mammals, they nurse their young with milk, have hair, are warm-blooded, and belong to a larger group of animals called vertebrates (see *How Are Whales Like Other Mammals?* and Appendix 1). Vertebrates are animals with segmented backbones. Characteristically whales have slender, torpedo-shaped bodies with projecting forelimbs, or flippers, but without hind limbs. Because of their complete adaptation to life and locomotion in the water, they have assumed a shape that looks very much like that of fishes. Whales bear no closer relationship to fishes than we do. There once was a time in the English language when *fish* referred to any creature that dwells in the sea and was extended to whales as well as dolphins, porpoises, and turtles. With the expansion of public education in the latter part of the nineteenth century, particularly with Thomas Henry Huxley's efforts to educate the working man, people became aware of the differences in marine animals and began to use more specific terms. Nowadays there is still some uncertainty about what people mean when they use the term *whale*. Do they mean only the "great whales" or any member of the mammalian order Cetacea?

WHAT ARE CETACEANS?

Cetacean is a technical term for members of the mammalian order Cetacea (whales, dolphins, and porpoises). The term is derived from the Greek root *ketos*, meaning

A group of humpbacks cooperatively lunge feeding.

"any large sea animal or sea monster." Cetaceans are totally adapted to the aquatic environment. They are incapable of moving about on the land; they eat, sleep, and reproduce (including giving birth) in the water. Only one other order of mammals is comparably adapted to the marine environment, and that is the order Sirenia, which includes manatees, dugongs, and sea cows. Sirenians were known formerly as herbivorous cetaceans. The pinnipeds, another group of marine mammals that include seals, seal lions, and walruses, must return to land (or ice) to give birth to their young.

Most cetaceans live exclusively in saltwater; however, occasionally they venture into freshwater without damage. One exception is the group of cetaceans known as river dolphins, most of which live in freshwater throughout their lives.

WHAT ARE THE DIFFERENCES BETWEEN BALEEN AND TOOTHED WHALES?

The living whales are divided into two suborders, the baleen whales, or Mysticeti (from the Greek *mystax* = upper lip [mustache] + *ketos* = whale), and the toothed whales, or Odontoceti (from the Greek *odontos* = tooth + *ketos* = whale). Even though modern baleen whales do not have teeth as adults, they do have tooth buds as fetuses, which resorb before they are born.

Fundamentally, the difference between baleen whales and toothed whales is the presence or absence of teeth. Baleen whales have baleen, and toothed whales have teeth (see *How Many Teeth Do Whales Have?* and *What Is Baleen?*). These characteristics reflect a primary difference in feeding behavior. Baleen whales tend to feed on small animals that school and are capable of being engulfed many at a time. Toothed whales hunt and feed on individual prey items. Some of the early baleen whales found fossilized, however, had teeth and possibly baleen as well. Many anatomical differences are correlated to feeding, such as the structure of the skull. The skull bones that form the rostrum (nose, beak) in the baleen whales are strengthened with buttresses above and below the eye, whereas toothed whales have only the buttresses above the eye. These modifications have been referred to as telescoping of the skull (see *Why Do Whale Skulls Differ from Those of Other Mammals?*).

Differences between the two suborders also occur in their acoustic abilities. Toothed whales use active sonar (echolocation) and emit high frequency bursts of sound that reflect from prey. Baleen whales have not been shown to be capable of active echolocation.

Left: A young bowhead, taken by Eskimos, showing the long baleen plates attached to its upper jaw. The animal is lying on its back with its mouth open. *Right:* The open mouth of a captive killer whale showing its strong teeth. Such teeth are characteristic of a toothed whale.

WHAT ARE THE DIFFERENCES AMONG WHALES, DOLPHINS, AND PORPOISES?

The English common (vernacular) names "whale," "porpoise," and "dolphin" are applied in slightly different ways in various English-speaking countries. Size is the determining factor in whether a cetacean is called a whale. Basically, any cetacean so large that it cannot be moved readily by a small group of people is called a whale. All others are porpoises or dolphins. Thus the term *whale* is used for various distantly related cetaceans like pilot whales and killer whales, beaked whales and sperm whales, and the baleen whales.

In Europe the term *porpoise* is applied only to members of the small (in terms of both size and number of species) odontocete family Phocoenidae, the harbor porpoises and their kin. All other small toothed whales are called dolphins. This includes the river dolphins, Platanistidae, and the oceanic dolphins, Delphinidae. In this strict usage, several general differences become evident. Porpoises have a blunt forehead; they do not have the pronounced "beak" of most dolphins. Porpoises also have shorter, spoon-shaped (spatulate) teeth. Dolphins have slender needlelike teeth.

In North America the terms *porpoise* and *dolphin* are used more or less interchangeably. The common species that is known as the bottlenose dolphin (*Tursiops truncatus*) in the Atlantic is also called the bottlenose porpoise in the Pacific. The only consistent North American usage is that the members of the family Pho-

A bottlenose dolphin showing its conical teeth. This bottlenose is opening its mouth in anticipation of food. Porpoises have much smaller teeth, which are spoon-shaped.

coenidae are always called porpoises. *Bottlenose* and *bottlenosed* are equally correct grammatically. In the early 1970s, however, scientists reached a "gentleman's agreement" to call this dolphin species bottlenose, not bottlenosed.

WHAT ARE THE MEANINGS OF CETACEAN NAMES?

Whale comes from the Anglo-Saxon *hwael,* meaning "a large fish." It is related to the Scandinavian *hval* and the German *Wal.* Some etymologists have supposedly traced it back to the root for *wheel* and liken the rolling motion of the whale to the turning of a wheel. Others trace *wheel* back to a separate Anglo-Saxon word, *hweol.*

Porpoise is derived from the Latin phrase *porcus* (hog) + *piscis* (fish), literally, a sea pig. This relationship to *pig* is reflected in many vernacular names for the porpoise, for example, puffing pig (Newfoundland), *chancho marino* (South American Spanish = sea pig), *Meerschwein* (German = sea pig).

Dolphin can be traced back to the Greek *delphin* = common dolphin, *Delphinus delphis.*

WHAT ARE RORQUALS?

Rorquals are baleen whales of the genus *Balaenoptera*, which includes the blue, fin, sei, Bryde's, and minke whales. These whales are equipped with abundant distinctive grooves that begin on the lower (ventral) surface of their throats and continue onto their chests. The only other member of the family Balaenopteridae and the only other whale with numerous ventral grooves is the humpback, but it is normally not included as a rorqual (see *What Is the Purpose of the Throat Grooves?*). The term *rorqual* as it was commonly used by early whalers referred to those whales that were either too swift or too difficult to kill, those that sank after death, or those that failed to yield enough baleen or oil to make it worthwhile for the whalers to pursue them (see *Why Did Whaling Decline?* and *When Did Whaling Decline?*).

Rorqual is known to be a Scandinavian word, but beyond that its origin is uncertain. The last part of it certainly comes from *hval*, meaning "whale." The first part of the word could come from the root *ror*, meaning "reed" and presumably referring to the ventral grooves, or from *reydhr*, meaning "red."

WHAT IS THE STORY OF THE MINKE WHALE?

Whaling lore tells us how the minke whale got its name. As the story goes, a man named Miencke was one of the crewmen on the *Spes and Fides* (Hope and Faith), the vessel of Svend Foyn, the Norwegian who started modern whaling in the mid-1800s. All the members of the crew took turns on lookout in the crow's nest. The principal target of the early whalers was the blue whale, followed by the fin. The other rorquals were not considered worth their time. The crewman Miencke had difficulty in differentiating the rorquals by the height of the blow. After the crew had chased down several whales that Miencke had spotted and misidentified, they came to call the lesser rorquals Miencke's whales. As time passed, even the lesser rorquals became prey, and the term *Miencke's whale* was left to the smallest, the little piked whale, or, as we know it today, the minke whale.

WHAT ARE BEAKED WHALES?

Beaked whales are toothed whales and members of the family Ziphiidae. Early in their evolutionary history, beaked whales separated from the main toothed-whale lineage, and the relationship of the two groups is still unclear. Beaked whales are distinguished by a reduced number of teeth, so that in most species a single pair of teeth occur in the lower jaw and no teeth occur in the upper jaw. In some species of the genus *Mesoplodon* the lower teeth are dramatically enlarged to form tusks, which are probably used by males in fighting. Indeed, males do fight a lot, and their

A bottlenose whale showing its characteristic forehead and beak during a blow. Although dolphins have similar structures, beaked whales are the only whales that have a beak.

bodies are frequently covered with tooth scars (see *Do Whales Fight?*). One other peculiarity of beaked whales is their snout, which is prolonged into a "beak." In the male of most species this snout, or rostrum, becomes extremely hard and dense as the whale ages. Studies have shown that it is the densest animal tissue known, denser than elephant ivory and, in some cases, as dense as the mineral quartz. Speculation is that the increased density of the rostrum may play a part in combat (like bartenders' old practice of putting lead in pool cues to increase the weight for using them as weapons in bar fights). Increased rostral density also alters the sounds emitted by the male.

Beaked whales live in the open ocean. Because their blow is projected forward and not upward, they are seldom seen, and very little is known of their natural history. Specimens of some species are rare. In fact, one beaked whale, Longman's beaked whale, *Mesoplodon pacificus*, is known only from two skeletons; it has never been seen alive.

WHAT IS A BLACKFISH?

"Blackfish" is a common name that was given by fishermen to small whales with essentially black coloration. Many species were given this name; the most common one is the pilot whale (*Globicephala melas* or G. *macrorhynchus*). Blackfish Creek on

Cape Cod is a famous locality where local fishermen used to drive pilot whales ashore. Other species that were commonly called blackfish are the false killer whale (*Pseudorca crassidens*), Risso's dolphin (*Grampus griseus*), some less common, smaller, offshore dolphins, the pygmy killer whale (*Feresa attenuata*), and the melon-headed whale or many-toothed blackfish (*Peponocephala electra*).

HOW MANY SPECIES OF WHALES EXIST TODAY?

There are 86 species of whales, porpoises, and dolphins living today. We must bear in mind that the classification of cetaceans is fluid. Over the years, 656 species names have been applied to whales and dolphins. The 86 species that we recognize today belong to the families in Table 1.

HOW ARE WHALES LIKE OTHER MAMMALS?

Whales provide their young with milk, as all other mammals do. Whale milk is produced by specialized skin glands (mammary glands) hidden within mammary slits

One of the many species known as blackfish, a captive short-finned pilot whale opening its mouth to get food.

TABLE 1. SUBORDERS AND FAMILIES OF THE ORDER CETACEA

Mysticeti (baleen whales)—14 species
    Balaenidae (right whales)—4 species
    Balaenopteridae (blue whale, fin whales and kin, humpback whale)—8 species
    Eschrichtiidae (gray whale)—1 species
    Neobalaenidae (pygmy right whale)—1 species
Odontoceti (toothed whales)—72 species
    Delphinidae (oceanic dolphins)—36 species
    Monodontidae (beluga and narwhal)—2 species
    Phocoenidae (porpoises)—6 species
    Physeteridae (sperm whales)—3 species
    Platanistidae (river dolphins)—5 species
    Ziphiidae (beaked whales)—20 species

located between what would be hind legs if whales were to have hind limbs. Mammals can develop mammary glands anywhere along the "mammary line," which extends from between the front limbs to the hip area (the inguinal position). The term *mammals*, in fact, refers to the mammary glands. The German word for mammals, *Saügethiere*, means "sucking beasts." Whales, like other mammals, give birth to live young.

Surprisingly, whales have hair, another characteristic shared by all mammals. In whales, the hair is reduced to a line of hair follicles on the head. These follicles are in the same position as the sensory hairs of terrestrial mammals, along both the upper and lower lips, the nose (or the blowhole), and the eyes. They are most noticeable in young animals but are found in all whales of any age. They also form the nucleus of the growths called callosities, which are the bumps on the head so characteristic of humpback and right whales.

Most important, whales have a lower jaw consisting of one bone, like the dentary, or bone that holds the teeth, of other vertebrates, and they have a chain of three bones (auditory ossicles) in the middle ear: the malleus, the incus, and the stapes (equating in the human ear to the hammer, the anvil, and the stirrup), one of which (the malleus) was originally in the lower jaw.

HOW ARE WHALES DIFFERENT FROM TERRESTRIAL MAMMALS?

Whales differ from terrestrial mammals in several ways. First, they have lost their hind limbs as a result of becoming fully aquatic. Other marine mammals such as seals have modified their hind limbs into hind flippers, which they use to propel themselves through the water and to maneuver on land. Whales use their flukes, a

A southern right whale showing the flukes that are characteristic of all cetaceans. These animals sometimes hold this pose for several minutes.

modification of their tails, for propulsion. The only other mammal group that has completely lost its hind limbs is the sea cows, which are also completely aquatic.

Second, because of the complexities of hearing under water as opposed to hearing in air, the bone that houses the middle ear (tympanic bulla) in whales has become modified. The bulla is enlarged and consists of thicker, denser bone than that of terrestrial mammals. Whales have developed a unique feature, the sigmoid process, in connection with the modification of the eardrum. Instead of a tight, flat membrane, the eardrum is prolonged laterally into a shape that has been called the glove finger. In large baleen whales, the eardrum looks remarkably like a finger in a rubber glove. (See *How Can Cetaceans Hear in Water?*)

Last, all whales have lost any trace of the external ear. This characteristic is shared by certain other aquatic mammals (true seals and sea cows) and some burrowing mammals (moles and naked mole rats).

HOW DOES THE GENETICS OF WHALES COMPARE WITH THAT OF OTHER MAMMALS?

Like other living things, cetacean genetic information is carried on chromosomes in the nucleus of the cell. The first foray into comparative genetics of whales consisted of chromosome counts. Other mammals show considerable variation in the exact number of chromosomes (karyotype). For instance, house mice can have different chromosome counts if they live in different buildings. Early work with whales, however, demonstrated an amazing uniformity in chromosome structure and karyotype. Whales have a chromosome count of 44, except for the sperm whales (both giant and pygmy) and the beaked whales, which have 42. The sperm and beaked whale families of toothed whales separated from the others very early, and it is not entirely surprising that they do not retain what appears to be the primitive cetacean chromosome count of 44.

Today, the amount of information that one can obtain from genetic sampling is impressive. First, this technique can confirm the sex of individual whales. Biologists who perform behavioral studies of cetacean populations recognize individuals by the shape of the dorsal fin, pattern of pigmentation, and scarring. Unless they are extremely lucky and see the genital region of the animal's ventral surface, they cannot identify whether it is male or female. Biopsy darts now obtain samples that can reveal the sex of the individual, as well as the pesticide residues in the blubber. In addition, DNA from the sample can be sequenced to tell what genetic population the individual is from, and, in certain cases, it can determine the parentage of the individual.

WHAT DOES THE DNA REVEAL ABOUT THE ANCESTRY OF CETACEANS?

DNA sequences confirm that the ancestors of cetaceans came from a primitive group of mammals called condylarths, which were most closely related to artiodactyl (even-toed) ungulates (cows, sheep, and antelopes, but not horses, which are perissodactyl [odd-toed] ungulates). They also confirm that baleen whales and toothed whales are very different. Recently some DNA studies show a very close relationship between whales and hippopotamuses (which are also artiodactyls).

A bit of a stir was also raised recently (1993) by a cladistic analysis based on DNA sequences that indicated that sperm whales, which are toothed, are more closely re-

Left: A small biopsy dart that has just struck a humpback in southeastern Alaska. The tip of the dart is designed to penetrate the skin, collect a sample, and then fall off to be retrieved by the researchers. *Right*: A biopsy dart that has just been used to collect a sample of a whale's skin. You can see the blubber protruding from the tip of dart.

lated to baleen whales than they are to the other toothed whales. One must bear in mind that all of these analyses are based on statistics, and the picture of relationships given by one set of statistical assumptions may be very different from that given by another. We still are not sure of the precise relationships of the sperm and beaked whales to other cetacean families, both living and fossil.

WHERE DO WHALES AND DOLPHINS LIVE?

All of the great whales live in saltwater, in the Arctic, Antarctic, Atlantic, Pacific, and Indian Oceans of the Northern and Southern Hemispheres. Even though whales and dolphins are able to survive in freshwater and there is enough room and food for them in some of the larger lakes, the entry to those lakes involves rivers where limited space is available for a great whale to maneuver. In addition, the surface of lakes that exist in temperate climates occasionally freezes over, which would prevent cetaceans from breathing.

Bodies of water smaller than oceans are called seas, bays, gulfs, and sounds. Some are sufficiently isolated to have developed distinct genetic populations of whales and dolphins. Such is the case with harbor porpoises and bottlenose dolphins in the Black Sea, which is connected to the Mediterranean Sea via the Dardanelles, the Sea of Marmara, and the Bosporus. Populations of cetaceans live in every sea but the Caspian, which no longer has a connection with the marine environment. Seals live in the Caspian Sea but no whales.

Some coastal porpoises, dolphins, and whales occasionally enter rivers for short periods of time without physiological distress. They do not acclimate to freshwater the same way that marine fishes must; they simply breathe in air through their blowholes. The breathing of fishes, in contrast, involves a closer relationship with the water, because they take water in through the gills, where oxygen is extracted. In essence one could say that fishes "breathe water," whales and dolphins "breathe air." One disadvantage for dolphins is that prolonged contact with freshwater may have a softening effect on their skin and may make the animal more susceptible to fungal skin diseases.

River dolphins (Platanistidae), a separate family from oceanic dolphins (Delphinidae), live their entire lives in freshwater rivers and lakes. However, some oceanic dolphins, the Irrawaddy River dolphin of Southeast Asia, and the tucuxi of South America also live their entire lives in freshwater.

DO ALL WHALES MIGRATE?

Most whales migrate, but not all of them do. The bulk of our information about whales is derived from commercial whalers who sought out migratory populations of whales. These populations make up the highest concentrations and today are the most sought after by whale watchers. For many years, people thought they knew the migration pattern of the California gray whale. Biologists thought that the whole population moved more or less as a unit from their breeding grounds in the lagoons of Baja California to the feeding grounds in the Bering Sea. Recently, it was discovered that some gray whales do not complete the entire migration but instead stop for the summer along the way. The population of gray whales that summer off Vancouver Island is one of these.

The existence of year-round whale residents is still open to debate. A resident population of Bryde's whales may occur off the coast of South Africa, and some of the whales that frequent the area around the Costa Rican Dome may be year-round residents.

WHY DO WHALES MIGRATE?

Whales migrate to take advantage of the summer feeding grounds in polar waters and return to more equatorial waters in the winter to breed. Oceanic circulation causes areas of "upwelling," where nutrient-rich waters are forced up from the deep ocean to the surface, forming the basis for enormous blooms of plankton and small fishes. One such area along the Humboldt Current off western South America supported, in the early 1970s, the largest fishery in the world, the Peruvian anchoveta

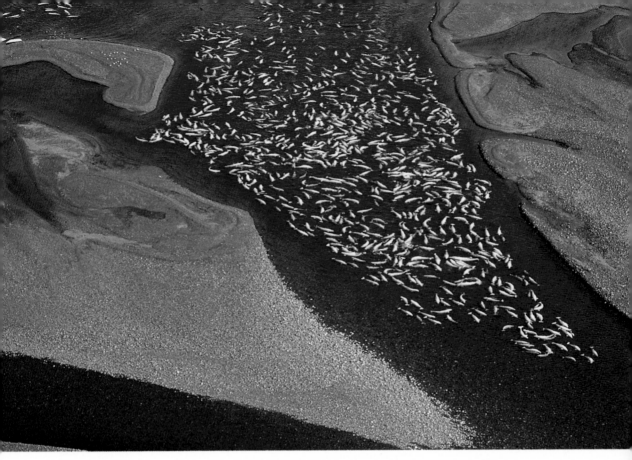

Belugas feeding in the shallow waters of the Arctic. They congregate in thousands to form migratory herds.

fishery. Unfortunately the fishery was overexploited, so it collapsed. Other upwelling areas occur in a belt that surrounds Antarctica and in areas of the North Atlantic and North Pacific. Early 1900 whaling records demonstrate that fin and blue whales captured by the whalers early in the season when they had just arrived in Antarctic waters were extremely thin. Whales taken at the close of the summer season were extremely fat.

WHERE AND WHEN DO WHALES MIGRATE?

Migration, in the sense of seasonal movements of an entire whale population, can be latitudinal (north-south), or onshore-offshore, or a combination of both. We now have data on the migratory routes of the whales that were taken commercially in the 1900s and of certain populations of whales that have been extensively studied, such as humpback whales in the North Pacific and North Atlantic, California gray whales, and right whales in the North Atlantic.

In general, whales migrate toward the poles in the summer (in both the Northern and Southern Hemispheres) and toward the equator in their respective winters.

This means that in April, North Atlantic humpback whales are migrating from their wintering grounds in the Caribbean toward their summering grounds off the coasts of Newfoundland and Iceland. At the same time, South Atlantic humpback whales are headed from their summering grounds in the Antarctic to their wintering grounds along the coasts of South America and Africa. Whales tend to migrate north in March and April and south in September and October (regardless of the hemisphere they live in). This means that it is extremely unlikely that the populations of migratory whales in one hemisphere will ever encounter the population in the other hemisphere, even though they may share equatorial breeding grounds. However, a humpback migrating from the Antarctic Peninsula to the Caribbean Sea, off the northern coast of Colombia, has been documented. This humpback covered at least 8,334 kilometers one way.

The migratory routes of whales are not rigid. With a closer look at how whales behave, we should be able to account for challenges to the norm. For example, the traditional belief used to be that all gray whales migrated to the Bering Sea in the summer. Now, we find that certain populations of gray whales spend the summer in British Columbia (see *Do All Whales Migrate?*).

HOW FAR DO WHALES TRAVEL? WHAT ARE THEIR SPECIFIC ROUTES?

California gray whales routinely migrate from calving areas in Baja California to the Bering and Chukchi Seas and back—18,000 kilometers round trip.

One humpback was sighted along the Antarctic Peninsula (64°20′ south latitude, 62°27′ west longitude) on 19 April 1986 and was resighted off the coast of Colombia (2°57′ north latitude, 78°12′ west longitude) on 28 August 1986, a straight-line distance of 8,334 kilometers. Normally the longest humpback migration is from the West Indies to Jan Mayen Island (6,435 kilometers) or to Bear Island (7,940 kilometers), both north of Norway.

WHAT ARE THE HAZARDS OF WHALE MIGRATION?

The hazards of migration relate to the seasons during which whales migrate, the relative lack of food on the migratory route, and the presence of newborn animals. Seasonal storms come up, which affect both the adults and the calves; food is scarce in certain areas; and the migration may subject the animals to human hunting. These migratory hazards do not constitute serious hurdles for the population as a whole.

Whales migrate in the spring and fall months, when the weather tends to be nasty in many parts of the world. Because of this strong migratory compulsion, they expose themselves to the vicissitudes of the weather, and the weaker animals may die. This is especially true of young (newborn) animals. The strong survive.

DO WHALES FEED WHEN MIGRATING?

Whales, like any other animal, feed when food is present and when they are hungry. The notion that whales do not feed during migration came from early whalers' observations of relatively emaciated whales that had arrived at the summer (feeding) grounds. However, there has been little opportunity to observe whales during their migration, so we are still not certain about the answer to this question. We have ample data, though, indicating that gray whales do not feed in the vicinity of San Francisco Bay during migration.

HOW DO SCIENTISTS TRACK WHALES AND DOLPHINS?

Scientists have used several different methods for tracking whales, including visual sightings, tags, and passive acoustic techniques. As anyone who has gone out whale watching can attest, it can be extremely difficult to follow whales visually. Add to this the problem of keeping track of whales during the night or for long periods of time, and we can see why the data derived from visual tracking are limited. Visual identification is very important in monitoring whale behavior at the endpoints of migration but not in determining the route of migration itself.

During the 1920s and 1930s, when whale movements were of extreme economic concern to whalers, much attention was given to tags. Through trial and error, the Discovery tag was developed. It was named for the Discovery Committee, a team of British scientists who were funded from taxes on the whaling industry to investigate whale biology. The Discovery tag, consisting of a short piece of stainless steel tubing with information engraved on it, was fired from a 12-gauge shotgun into the side of a whale, where it penetrated the muscle and became encapsulated by tissue. After a whale was captured by a whaling company, the tag might be discovered either when the whale was flensed (see the glossary) or when its fat was rendered. The tag, with time of capture and locality data added, was turned over to scientists who kept a file system on where and when the tag originally was shot into the whale. That provided two endpoints of the whale's movements. Most of the tags were picked up in the same areas where the whales were first tagged, but some com-

A satellite tag just after attachment to a Canadian beluga. In an attempt to monitor this beluga's migration, scientists will record the signals from the transmitter on the tag after they are picked up by a satellite, which will determine the precise location of the animal.

A tagged beluga with a school during migration. It is very important that a tag does not cause an animal to modify its habits.

mercial whaling occurred on the winter breeding grounds, so that the tags collected there slowly enabled biologists to get a general idea of population movements.

Because the Discovery tag was not visible on the outside of a whale, it was essentially worthless until the whale was killed. As a result, greater efforts were made to fashion tags that were visible. Some people tried attaching streamers to the tags, but this was not successful. For a period, people painted markings on whales to serve as tags, which resulted in the invention of a hand-held pressurized spray can that was capable of painting a number on the side of a whale. Freeze-branding numbers onto the dorsal fin of dolphins was also successful but limited to animals that could be captured and held immobile while the brand was applied.

With the advent of miniaturized electronics, it became possible to encapsulate waterproof radio transmitters and attach them to whales. These early simple "radio tags" provided a beacon by which the position of the whale or dolphin could be determined and plotted, but radio tags suffered from limited battery life and limited range. The whale or dolphin still had to be followed with a boat, and the lifetime of the tags was just a few weeks. Today, whales and dolphins are equipped with

microwave transmitters whose signals are picked up by satellites and relayed to re-
ceivers that can be hundreds or thousands of kilometers away from the animal. The
transmitters have become more efficient, and the batteries have improved so much
that it is possible to collect data for several months. (See *How Do We Determine
What Whales Are Doing Underwater?*)

The preceding paragraphs dealt only with active radio tags, active in the sense
that they actively transmitted a signal that provided the data. However, whales are
somewhat noisy, so it should be possible to determine their location also by listening for them with hydrophones, a "passive" acoustic technique. At one time, biologists used this technique to determine the location of bowhead whales under the
ice in Alaska. By using a hydrophone array that permitted precise determination of
a whale's position when it vocalized, they were able to track the bowheads.

Passive acoustic techniques were also used to discover if there is a difference between the nighttime and daytime rates at which gray whales migrate. This research
took place in the 1960s in a project off the coast near San Diego, where gray whales
were recorded both at night and in the daytime. The data collected at that time
showed no difference in migratory rate.

However, more recently thermal sensors, which can pick up whales coming to
the surface and blowing, were used on one of the survey sites along the California
coast to monitor gray whales' nighttime migration. This study found a rate of whale
movement during the night up to 39 percent greater than during the day.

With the decline of the Soviet Union as a potential threat, the U.S. Navy recently (1993) made available its vast low-frequency hydrophone network (sosus,
or *sound surveillance system*), with which it monitored the position of ships by listening to the noise made by their propellers. Biologists found that with this array
they could track whales by the noises they emit. Some of the whale vocalizations
were sufficiently diagnostic to permit tracking of individuals. A blue whale was followed for 43 days over a course of 2,700 kilometers—a record for passive acoustic
tracking. Unfortunately, this kind of network is extremely expensive to maintain;
therefore, unless the technology changes, we may see its demise in future years.

HOW MANY WHALES ARE THERE?

When counting whales began, it became immediately evident that, to understand
what a population of whales is doing and to come up with a reliable count, it is necessary to recognize individuals. In addition, data need to be established on both the
short- and long-term life history and movements of individuals in order to derive
theories about the activities of a population.

The first successful attempts at this data gathering were based on implanted tags

A humpback showing its flukes, which have a very distinct pattern of pigmentation. Humpbacks commonly display their flukes when preparing for a dive. This gives biologists a chance to photograph them to keep records of the identity of the whales.

(Discovery tags), radio tags, and satellite tags (see *How Do Scientists Track Whales and Dolphins?*). As whaling ceased to be economically and ethically viable, researchers looked for other ways to recognize individuals. The first large-scale individual recognition project was the humpback whale fluke catalog that was undertaken by researchers at the College of the Atlantic, who formed the Gulf of Maine Whale Sighting Network, photographing humpback flukes and documenting individuals on the basis of fluke pigmentation patterns. In 1977 they published the first identification catalog. Many individuals and organizations have contributed to the project since its inception. Basically, it has involved photographing the flukes of humpbacks diving and using the complex pigmentation on the underside of the flukes as a way of "fingerprinting" the whales. Researchers recorded the time and place where any particular whale was photographed and sent that to a central data bank. When the whale was seen again, it was photographed, and the date and position were recorded. Over the years this has resulted in a large accumulation of data that were not limited to the time of tagging and the time of death,

Researchers comparing photos of flukes with those kept in enormous photo archives. Through the use of fluke catalogs, researchers are able to identify individual whales and compile records that show where and in whose company the whales were on a specific day.

as they were in Discovery tagging. The newer technique enabled researchers to determine the precise destinations of whales and to establish ties between the summer feeding grounds and the winter breeding grounds of individual humpbacks.

Techniques of individual visual recognition and documentation have been applied to almost every species of whale and dolphin that is frequently seen, including right, blue, fin, bottlenose, gray, killer, Blainville's beaked, and sperm whales as well as Atlantic spotted, Amazon river, tucuxi, Indo-Pacific humpback, bottlenose, Hector's, Risso's, and spinner dolphins.

Recently a project known as YONAH (Years of the North Atlantic Humpback) was carried out. This project attempted to biopsy individually recognized humpbacks and use the biopsy samples to determine the relationships of the whales through the techniques of molecular genetics.

When a whale population has been subjected to sufficient observation, as has occurred with the North Atlantic humpback and the right whale, it is merely a matter of adding up the numbers of whales recognized to obtain a total population count. This, of course, is dependent on the completeness of the count, which is measured by the number of individual whales encountered before an unrecognized one turns up. That procedure gives us a number that is indicative of the unknown percentage of the population. For instance, if we sample 10 familiar individuals be-fore encountering 1 new individual, we are dealing with a population that is 90 per-cent known.

Problems occur in counts of whale populations that are less well known. In those cases we must use complex assumptions to determine the number of whales known (sampled, surveyed, tagged) and the percentage of the population that that figure rep-resents. This is a common situation in cetacean population estimates (see Table 2).

TABLE 2. WORLD POPULATION OF SELECTED WHALES

Common Name	Scientific Name	Population Estimate	
		Pre-exploitation	Present
Blue	*Balaenoptera musculus*	228,000	14,000
Fin	*Balaenoptera physalus*	548,000	120,000
Sei	*Balaenoptera borealis*	256,000	54,000
Bowhead	*Balaena mysticetus*	30,000	8,200[a]
Sperm	*Physeter catodon*	2,400,000	1,950,000
Northern right	*Eubalaena glacialis*	no estimate	300 western N. Atlantic[a]
Southern right	*Eubalaena australis*	100,000	3,000
Humpback	*Megaptera novaeangliae*	115,000	10,600 N. Atlantic[b] 6,000–8,000 N. Pacific
Gray	*Eschrichtius robustus*	20,000+	26,600 eastern N. Pacific[c] a few hundred western N. Pacific
Bryde's	*Balaenoptera edeni* [+ *brydei*]	100,000	90,000
Minke	*Balaenoptera acutorostrata*	140,000	725,000
Killer	*Orcinus orca*	no estimate	no estimate
Pygmy right	*Caperea marginata*	no estimate	no estimate
Narwhal	*Monodon monoceros*	no estimate	35,000
Beluga	*Delphinapterus leucas*	no estimate	50,000

General source: Adapted from "Whither the whales," *Oceanus* 32, no. 1 (spring 1989): 12–13.

[a]Data from Marine Mammal Commission, *Annual Report to Congress 1997*, 31 January 1998, Bethesda, Mary-land.

[b]Data from T. D. Smith et al., "An ocean-basin-wide mark-recapture study of the North Atlantic humpback whale (*Megaptera novaeangliae*)," *Marine Mammal Science* 15, no. 1 (1999): 1–32.

[c]Data from D. J. Rugh et al., *Status Review of the Eastern North Pacific Stock of Gray Whales*, NOAA Techni-cal Memorandum NMFS-AFSC-103, 1999. The estimate has been shown elsewhere as 22,600 (Le Boeuf 1998: 106).

WHEN DID PEOPLE BEGIN TO STUDY WHALE ANATOMY?

Aristotle (384–322 B.C.), a Greek natural philosopher, was the first author to treat cetaceans biologically. He was familiar with both great whales and dolphins, although it is impossible to tell how much of his knowledge came from first-hand observation and how much was based on conversations with fishermen. Nonetheless, he clearly understood that cetaceans were mammals—that they suckled their young, breathed air, and had hair. It is worth noting that his principal biological work, *Historia Animalium*, a Latin title, actually was written in Greek as *Zoia Istorion*. The only remains of the work are fragments of Latin copies.

Pliny the Elder (A.D. 23–79), a Roman naturalist who perished in an eruption of Vesuvius, was a natural history compiler. In the preface to his main work, *Naturalis Istoria*, he claims to have assembled 20,000 facts from 2,000 books written by 100 authors. These books that Pliny cites have disappeared, but they contained abundant natural history observations on whales and dolphins. This body of data indicates that the classical world was familiar with the basic biology of whales and dolphins.

Because of the difficulty in publishing books that had to be hand copied, essentially no natural history studies were made during the Middle Ages. With the development of printing using movable type in the middle of the fifteenth century (ca. 1450) and the publication of the famous Gutenberg Bible, this situation began to change. One of the first large compendia containing natural history observations was Olaus Magnus's principal work in 1555, *Historia de Gentibus Septentrionalibus Carumque Diversia Statibus, Conditionibus, Moribus, Ritibus, Superstitionibus* . . . (A history of the northern peoples and their diverse social states, conditions, customs, religious practices, superstitions . . .). Olaus Magnus wrote of many creatures that were still known from fables only, such as several types of great whales, the walrus, the narwhal, and the kraken (giant squid), which he portrayed as an enormous octopus.

Konrad Gesner, a German Swiss author, wrote his famous four-volume work, *Historia Animalium*, between 1551 and 1558. His books were widely circulated, and he is commonly recognized as the father of modern natural history. His illustrations, particularly those of whales, were widely copied. Like Olaus Magnus, he also wrote about many animals that were legendary, like the manticore and the unicorn. He was obviously familiar with the narwhal tusk, because he pictured the unicorn (*Einhorn*, as it is called in German) as having the head and body of a horse, cloven hooves, and a properly spiraled (sinistral, or left-hand, spiral) narwhal's tusk emerging from its forehead.

The sixteenth century produced the first authors whose first-hand experiences started to fill libraries. Two ichthyologists, Guilliame Rondelet and Pierre Belon, wrote monographs on fish anatomy in which they included cetaceans. Edward Tyson, an English anatomist, was the first to treat the anatomy of a cetacean in a work devoted solely to that purpose.

More than 100 years later in 1787, John Hunter presented his paper "Observations on the Structure and Oeconomy of Whales" to the Royal Society of London. Hunter's work brought the science of cetacean anatomy and natural history to its present state. It is interesting to note the changes in meaning of the word *oeconomy* (economy). In Hunter's day it was used for the field of knowledge that we now call ecology. We authors think that, if the English were updated, Hunter's paper would make an excellent survey of the generalities of cetacean biology.

HOW LARGE ARE WHALES?

There is a great diversity of size among whales, including both their length and their weight. The smallest is the dwarf sperm whale, *Kogia sima*. It reaches a maximum length of 310 centimeters and a weight of about 400 kilograms. The longest and heaviest whale is the blue whale, *Balaenoptera musculus*. The longest blue whale that was reliably measured was a female 28.5 meters long, taken by whalers at South Georgia Island, between Antarctica and South America. The heaviest blue whale ever weighed was a female 27.6 meters long at 190 metric tons. This animal was weighed in pieces, and most of the blood and body fluids were lost. Using the rule that, on average, blood and body fluids represent 10 percent of the dry weight of a vertebrate, we calculate her approximate total weight as 209 metric tons.

Right whales and humpback whales are much stouter than blue whales but not as long. A 17.1 meter male North Pacific right whale weighed 67.24 metric tons. This was a dry weight, so his approximate total weight was 74 metric tons.

A pair of stranded male sperm whales were weighed intact in the Netherlands in 1937. The smallest was 16 meters long and weighed 39 metric tons; the largest was 18 meters long and weighed 53 metric tons. With such enormous weights involved and with the limited availability of scales that could perform this task, it is no surprise that few records of whale weights exist.

HOW ARE WHALES AND DOLPHINS ADAPTED FOR LIFE IN THE WATER?

The primary adaptation of whales and dolphins for life in the water is the transformation of body form from that of a terrestrial creature to one that can survive in the marine environment. Cetaceans live in a medium (water) that is thicker than air, so they have become streamlined like most fishes and have lost projecting hind

A blue whale, seen from the back, that is just about to open its blowhole to take a breath. Baleen whales have paired blowholes, which can be seen on either side of the middle furrow on top of the head. The broad flat expanse of the blue whale's head is just about to show.

limbs. The shape of the head, the way it is attached to the body, the lack of constriction of the neck, and the shape of the forelimbs enhance the streamlining effect.

For whales and dolphins, hair is of limited importance as insulation in the water, so they have eliminated body hair except for the few hair follicles on the head (see *How Are Whales Like Other Mammals?*). In contrast, seals and sea otters retain hair for insulation in the air, because they are amphibious animals, spending most of their time on the water's surface or on land. Also, fur seals and sea otters trap air in their hair for insulation in the water. The lack of a hairy covering on cetaceans allows them to take advantage of the greater heat-conducting properties of water and provides them with the ability to give off heat when they are active. These two systems represent different evolutionary answers to the same problem.

An air-breathing animal living in the water must solve several problems. One is how to adjust the position of the mouth and nose. If you have ever been in a swimming race, you know how difficult it is to position your head so you can breathe. The nose of cetaceans is positioned on the top of the head so that they can breathe without undue effort. The nostrils of cetaceans are called blowholes. In baleen whales the blowholes are paired, lying on either side of a middle furrow; in toothed whales the blowhole is single. Toothed whales have developed an external nasal passage that is a complex of soft tissue overlying the paired bony exit from the skull and fusing into a single blowhole, whereas baleen whales do not have that specialization. In most toothed whales the blowhole is shaped with the horns of the crescent pointing forward. However, in the beaked whale genus (*Berardius*) and the dwarf and pygmy sperm whales (*Kogia*), the horns point toward the rear. Two more exceptions to this rule among toothed cetaceans are found in the river dolphins: the blowhole of the Yangtze river dolphin, or baiji (*Lipotes vexillifer*), is an oval, and the blowhole of the Ganges river dolphin (*Platanista*) is a longitudinal slit.

WHAT IS THE PURPOSE OF THE THROAT GROOVES?

Throat grooves appear in several families of whales. Beaked whales, gray whales, and sperm whales have one or two pairs of throat grooves that are relatively short and shallow. As far as we know the throat grooves in these families function in a limited manner to increase the capacity of the mouth in feeding. A growing body of information indicates that most, if not all, whales are suction feeders.

The whale family that has developed the most pronounced and extensive throat grooves is the balaenopterids—that is, the rorquals (fin whales and their kin) and the humpback. There, the grooves are many, from 14 to 100, deep (about 5 centimeters), and long, from the tip of the lower jaw to the navel in most species. They have become known as ventral grooves instead of throat grooves in rorquals because of the extension almost to the navel. They enable the mouth to expand its capacity many times over. Most balaenopterids feed by taking enormous gulps of water, then closing their mouth and expelling the water through their baleen, trapping fish and krill. The ventral grooves also appear to be covered with a vast network of blood vessels (vascularization), so they may also function as a radiator to dissipate heat.

WHAT IS THE TEXTURE OF CETACEAN SKIN?

Whales and dolphins have extremely smooth skin. Touching a whale or a dolphin is like touching an inner tube or a balloon full of air. The skin is so smooth and soft that it has been compared to the texture of a diver's wet suit. Cetacean skin normally feels cool to the touch but may feel warm in whales that have produced a lot of heat through exercise.

Head-on view of a beluga underwater, showing its smooth soft skin.

DO WHALES AND DOLPHINS HAVE HANDS AND FEET?

Like humans, whales and dolphins have appendages. In cetaceans they are the flippers, the flukes, and a dorsal fin. Of these, only the flippers, or forelimbs, are shared with land mammals (in humans they are known as arms). Whales and dolphins do not normally have legs, or hind limbs. The dorsal fin and flukes (or the tail fin) are features that changed as these animals evolved from their terrestrial ancestors and adapted to life in the water.

The forelimbs, or flippers, of whales and dolphins are modified into single functional units and vary in size and shape among the species. Although whales and dolphins have the same bony skeletal elements as terrestrial mammals, these elements are not visible externally. Internal support for the flippers is formed by the scapula, or shoulder blade, but there is no external sign of it, in contrast with most terrestrial mammals. Whales and dolphins have a shoulder joint, which permits movement of the humerus, or upper arm bone, on the scapula, but they do not have externally visible elbow or wrist joints. The bones are not fused together but are attached by a heavy fibrous tissue that permits only very slight movements. Humans and many other mammals still possess a clavicle, or collar bone, the supporting element that runs between the scapula and ventral part of the trunk, called the sternum. Cetaceans and a number of other mammals have lost the clavicle.

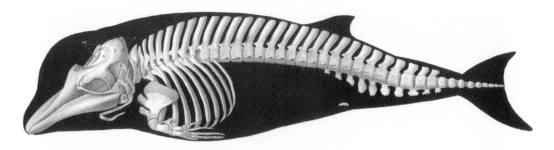

Skeleton with outline drawing of a bottlenose whale fetus, particularly showing the distribution of the bones in the fins (after Eschricht 1869).

The first element of the external forelimb, or flipper, is the humerus, known as the upper arm bone. It is relatively short in whales and dolphins, accounting for less than one-fifth of the length of the upper limb. In humans, the humerus is nearly half of the length of the arm.

The second element is composed of the radius and ulna, or the lower arm bones. The radius and ulna articulate with the humerus at the elbow joint. A process on the ulna (the olecranon process) forms the "funny bone," which causes so much pain in humans when it is hit (actually, it is the ulnar nerve that causes the pain). The radius and ulna in cetaceans form about a third of the length of the flipper, which is roughly comparable to humans, in whom these bones form slightly less than half of the length of the arm.

The next, or third, element is the manus, or hand. This consists of the carpal bones (or wrist bones), the metacarpal bones, and the phalanges (or finger bones). These bones are fused into the end of the flipper and are not capable of independent movement. The hand of whales forms about one-half of the length of the forelimb, as opposed to about one-quarter of the length in humans. This increased length in whales is principally composed of fingers.

The hind limbs of whales are usually present only in the vestigial pelvic girdle (the hip). The pelvic girdle, which supports the hind limbs in vertebrates, consists of three bones: the ilium, the ischium, and the pubis. These fuse early in the life of the animal and form the hip girdle. In whales, where the hind limb has been lost, the pelvic girdle remains as an attachment for some of the muscles concerned with reproduction. Scientists have not been able to determine how many of the three bones are represented in the cetacean pelvic rudiment.

Although the projecting hind limb is absent in most whales, it is sometimes present as a genetic anomaly in individuals. The hind limb and associated musculature have been described in humpback and sperm whales and in striped dolphins. A study of the incidence of the vestigial femur (the thigh) in Antarctic minke whales revealed that the femur was present in 25 percent of the whales sampled (in 13 out of 51 specimens).

The flukes of whales are the main organ of propulsion. They are lateral outgrowths of skin, blubber, and connective tissue at the end of the tail. They develop early in the life of the embryo as buds from the side of the tail. They grow out to the side and, in most whales, grow backward as well. This results in flukes whose hind margin extends beyond the bony tail, forming a fluke notch in the center. The flukes have a core of caudal vertebrae that extend back to the fluke notch. The beaked whales are the only cetaceans that do not have a fluke notch. Their vertebrae extend to the end of the tail, but the flukes have not developed backward to the extent that they form a notch.

HOW MANY FINGERS DO WHALES HAVE?

In most whales, the number of fingers has decreased from five to four, with the loss of the thumb, but the number of bones per finger has increased. Humans have a maximum of 3 bones per finger, but some dolphins, such as the pilot whale, have up to 14 bones in their middle fingers. The humpback has the longest flipper in the world, reaching up to 5 meters in length, or one-third of the length of the entire whale.

WHY DO SOME CETACEANS HAVE A DORSAL FIN?

The dorsal, or back, fin functions to stabilize cetaceans much as a keel on a sailboat stabilizes the boat. However, the role of the dorsal fin is not critical, because some species lack the appendage altogether (see *Why Do Some Cetaceans Lack a Dorsal Fin?*). The flukes, or tail fin, are flat horizontal structures that form at the end of the tailstock. They provide increased area for the propulsive movements of the tail, similar to the way rubber fins help a skin diver swim. The dorsal fin and the flukes do not have any bones to support them. They are composed of blubber and connective tissue.

WHY DO SOME CETACEANS LACK A DORSAL FIN?

Most whales and dolphins have a dorsal, or back, fin; in some the dorsal fin has been reduced to a series of bumps, such as in the gray and sperm whales; in some it has been lost altogether, such as in the beluga, narwhal, bowhead, right whale, northern and southern right whale dolphins, and the finless porpoise (*Neopho-caena*). Because of the amount of diversity in the habits of these animals, a single explanation for the loss of fin is unlikely to suffice. A partial explanation for the lack of a dorsal fin in the Arctic species may be that this fin is apt to impede movement around ice. But the killer whale sometimes visits the Arctic, and it has the highest dorsal fin of all.

HOW MANY BONES DO WHALES AND DOLPHINS HAVE?

Whales and dolphins have a varying number of bones depending on both the species and the age of the individual. An adult might have one bone (center of ossification) that is actually composed of six different bones that fused as the animal grew. In young animals of the same species, we can see these six separate centers of ossification. We have counted the number of bones of both adult and young bottlenose dolphins, *Tursiops truncatus*, and found that an average adult has up to 240 separate centers of ossification (see Table 3). By contrast, a young animal has 622 separate centers of ossification.

TABLE 3. COMPARISON OF THE NUMBER OF BONES IN A BOTTLE-NOSE DOLPHIN, A BLUE WHALE, AND A HUMAN (ALL IN ADULT FORM)

	Bottlenose Dolphin	Blue Whale	Human
Skull (incl. hyoids)	51	51	51
Vertebrae	65	64	34
Chevrons	16	13[a]	0
Ribs (right and left)	28	32	24
Sternal ribs	16	0	20[b]
Sternebrae	4	1	6
Forelimb (right and left)			
Girdle	2	2	4
Limb	6	6	6
Hand	50	50	54
Hind limb (right and left)			
Girdle	2	2	6
Limb	0	0	6
Foot	0	0	52
Total	240	221	263

[a]Number of chevrons uncertain.

[b]Cartilaginous elements in humans.

WHY DO WHALE SKULLS DIFFER FROM THOSE OF OTHER MAMMALS?

In adapting to an aquatic environment, the skulls of whales have become extremely modified from those of other mammals. Most adaptations reflect the needs of breathing and eating in the water.

Millions of years ago, as whales became more adapted to life in the water, and, in particular, as they increased their swimming speed, the nose began to migrate from the front of the head until it opened on the dorsal surface. That made it much easier to surface and breathe without pausing to turn the head. Human swimmers have accomplished the same thing through use of a snorkel. Anybody who has used a snorkel realizes the advantage in not having to turn his head to lift it above the water to breathe. So the migration of the nose to the top of the head was one of the three major forces that caused changes in the position of the whale's skull bones.

The second major force was important to the whale's locomotion. It involved the expansion of the dorsal muscles attached to the back of the skull and the forward migration of both the associated muscles and bones. These forces caused the bones of the skull to overlap. Because the overlapping resembles the way a telescope closes, the process has been called telescoping.

The very different manner in which telescoping has evolved in baleen and toothed whales is one of the basic differences between the two groups. The difference lies in the way the maxilla (the upper jaw) and occipital expand to cover the cranium. In toothed whales the ascending process of the maxilla extends backward over the frontal bone to meet (or almost meet) the occipital. This ascending process broadly covers the dorsal surface of the frontal bone. In toothed whales there is no descending process of the maxilla. In baleen whales the maxilla extends backward to the cranium both dorsally and ventrally, resulting in both ascending and descending processes. The ascending process in baleen whales is narrow and does not cover the orbital wing of the frontal bone. The descending process (the orbital plate) projects ventrally to the orbital wing of the frontal bone. The dorsal part of

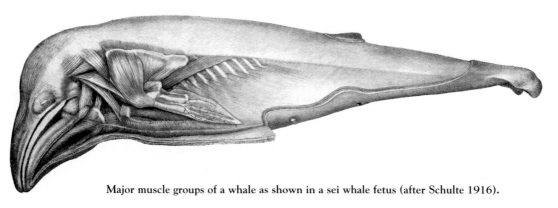

Major muscle groups of a whale as shown in a sei whale fetus (after Schulte 1916).

the occipital bone in baleen whales extends much farther forward than it does in toothed whales.

The third force that changed the skulls of cetaceans was the modification of their forelimbs into flippers, which deprived them of the ability to use these limbs to capture and manipulate food. Consequently, the snout became long and filled with teeth or baleen, resulting in extremes of form, such as the enormous jaw of the bowhead and sperm whale, which is nearly a third of the length of the whole animal.

HOW MANY TEETH DO WHALES HAVE?

Whales descended from terrestrial mammals that had a typical differentiated (heterodont) dentition; that is, their teeth were divided into incisors, canines, premolars, and molars. Early whales probably had the primitive mammalian complement of teeth: 3 incisors, 1 canine, 4 premolars, and 3 molars on each side of both the upper and lower jaws, making a total of 44 teeth. The teeth of living whales, however, have lost this differentiation and have become homodont. Whales use their teeth only for grabbing prey, which is swallowed whole; they do not chew their food.

The number of teeth in a cetacean species is variable. Some whale species have an increased number of teeth to make the capture of elusive, slippery prey easier. The total tooth count ranges from 2 in most of the beaked whales, to 76 to 100 in the bottlenose dolphins, 88 to 112 in the harbor porpoise, 160 to 200 in the common dolphin, and culminates in 210 to 242 in the franciscana (La Plata river dolphin). Sperm whales have only 36 to 50 teeth in their lower jaws, and none in their upper jaws, although the upper jaws do contain a variable number of vestigial teeth that do not erupt. Narwhals have two tooth buds in their upper jaws, and none in their lower jaws. The left tooth bud normally erupts and produces a tusk up to 3 meters long in the male. Females do not normally have an erupted tusk. Occasional reports of double-tusked individuals, both male and female, exist.

Cetaceans have just one set of teeth. There is some controversy over which set it is—deciduous (baby) or permanent. The full set of teeth is present at birth. Cetaceans do not add teeth with age.

Baleen is not related to teeth. Teeth and baleen are composed of entirely different structures (see *What Is Baleen?*).

WHICH WHALES HAVE TUSKS?

A tusk is a long pointed tooth, long enough to protrude from the mouth when the jaw is closed. The male narwhal has one upper tusk, which can project up to 3 meters. The other group of whales that have tusks are the beaked whales.

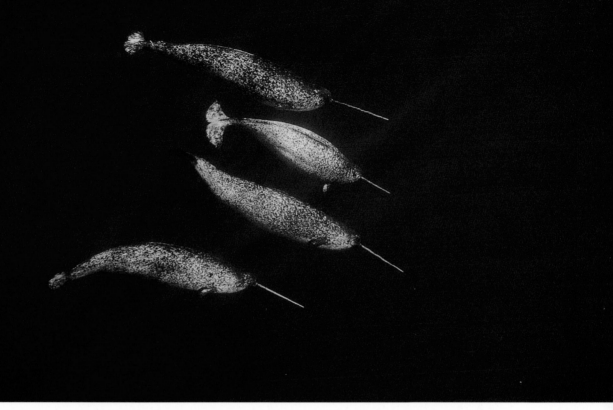

Four male narwhals, viewed from the air off Baffin Island, Canada, showing their exceedingly prominent tusks. The variable coloration of narwhals is seen in this photo, where one has light-colored flukes and the other three have dark flukes, which are more difficult to see.

Some beaked whales have projecting lower jaws that have teeth located at the tip, such as Cuvier's beaked, True's beaked, and Baird's beaked whales. However, most males of the genus *Mesoplodon* have markedly enlarged teeth correlated with enlargement of the jaw. These two enlargements cause the teeth to project as tusks.

WHAT IS BALEEN?

Baleen is a unique feature of baleen whales. It is organized into plates that hang down from the outer edges of the palate and are constantly regenerated. Baleen is composed of keratin, a protein that forms hair, fingernails, and horn. The baleen plates are composed of a line of bristles surrounded by a dense layer of keratin. In a continuous process, the keratin layer wears away on the inside surface (the surface toward the tongue), freeing the bristles. The bristles form a dense network that traps the prey the whale has taken into its mouth. Baleen is capable of being formed by the lining of the entire palate, as we have seen in a minke whale whose interior mouth had been injured. Perfectly functional baleen had formed on all of the scar tissue.

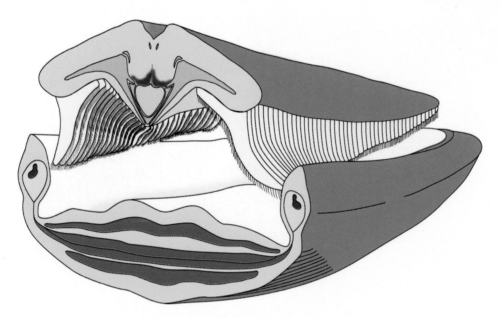

Cross-section of a minke whale head showing the orientation of the baleen (after Slijper 1962; Hentschel 1937; Hentschel 1912; and Delage 1886).

Baleen was one of the important elements in early commercial whaling. Before the development of spring steel and rubber in the middle of the 1800s, baleen was the material that was used for many elastic applications. The best baleen came from right whales. The baleen plates of the bowhead, formerly known as the Greenland right whale, are up to 4 meters long. Baleen was used for stays in corsets and collars, for buggy whips, and even for the springs in piano mechanisms (actions).

DO WHALES AND DOLPHINS HAVE SPECIALIZED HEARTS?

Whales and dolphins have four-chambered hearts, just as other mammals do; cetaceans' hearts have not become specialized in their evolution. However, the size of the heart is impressive in large whales. The largest heart was from a 27.6-meter blue whale, and it weighed 640 kilograms, which is as much as the weight of a small car. It was from an animal whose total weight was 190 metric tons, which means that the relative weight of the heart was 0.34 percent. However, this compares rather closely to the relative weight of the heart in humans, which is about 0.5 percent (0.42–0.66 percent); in cattle, 0.37 percent; in monkeys, 0.33 percent; and in cats, 0.45 percent.

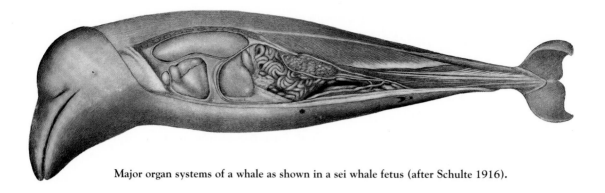

Major organ systems of a whale as shown in a sei whale fetus (after Schulte 1916).

WHAT KIND OF BLOOD VESSELS DO WHALES AND DOLPHINS HAVE?

Notwithstanding the lack of hind limbs and the circulation that serves them, whales and dolphins have a relatively normal mammalian circulatory system. However, some special blood vessel adaptations related to diving and living in cold water have occurred in cetaceans. An extensive arteriovenous network (*rete mirabile*, meaning "wonderful nets"), with its main vessels branching into a network of smaller ones, has developed through the chest, the spinal column, and the forelimbs. As a result, cetaceans lack a brachial artery (the main arterial supply to the forelimbs). Instead, they have a brachial *rete mirabile*, allowing for better control of the blood and making it possible to cut off the flow to the limbs during diving (see *How Can Whales Dive as Well as They Do?*).

HOW DO WHALE AND DOLPHIN LUNGS DIFFER FROM HUMAN LUNGS?

The respiratory system of whales and dolphins is similar to that of humans or any other mammal. Whales breathe through their nose (blowhole); the breath passes through the nasal passage of the skull and into the spout of the larynx (or voice box). The larynx is held by a muscular cuff (the nasopalatine sphincter) in the back of the nasal passage. When cetaceans are not eating, the connection between the nasal passage and larynx is maintained, but they can interrupt this connection and pull the larynx out of the nasal passage to allow food to pass. The larynx is connected to the lungs by means of a trachea and bronchi.

The lungs lie along the upper (dorsal) part of the chest cavity, as in all mammals. The lungs have several modifications from the normal mammalian pattern in terms of their fine structure. The lungs of terrestrial mammals are segmented; that is to say, they are divided into separate areas, each with its own blood supply. The lungs of cetaceans are not divided into separate segments, possibly because they are not exposed to the diseases and contaminants that ours are.

In humans and other mammals the trachea and larger bronchi are supplied with cartilaginous rings to keep them from collapsing. These rings end before the terminal sacs (the alveoli), where oxygen exchange between the air and the blood takes place. In cetaceans these rings continue down the respiratory tree to the borders of the alveoli. When whales and dolphins dive, the respiratory system starts to collapse. The cartilaginous rings, however, prevent collapse of the bronchi and bronchioles, allowing the alveoli to collapse first. Therefore, the residual air, with its load of nitrogen, is excluded from the areas where it would have contact with the blood. This is one factor that allows cetaceans to dive to great depths and not experience caisson disease (the bends) (see *Do Whales Get the "Bends"?*).

WHAT IS BLUBBER?

Blubber is known to most people as that portion of the whale that makes it fat and that made old-time whaling valuable. It is actually part of the skin of whales. Mammal skin is composed of two layers, the epidermis, or cuticle, which is the outer skin, and the dermis, or true skin (*cutis vera*). Blubber, or fat, is found in the dermis. The epidermis does not contain blood vessels or nerves, and in most mammals the outer layers are dried and hardened, forming a protective layer. If you take a fine sharp needle you can carefully insert it into the epidermis and feel no pain and experience no bleeding. Once you enter the dermis with the needle, the experience changes. The epidermis of humans is about 0.1 millimeter in thickness over most of the body. It can reach 1.4 millimeters in the soles of the feet. The epidermis in dolphins and porpoises is about 2 millimeters thick. It is thicker in large whales, reaching a reported maximum of 13.2 millimeters in right whales. The dermis is thicker, contains blood vessels, nerves, and, most important for whalers, fat. The dermis of a right whale can reach up to 700 millimeters in thickness. The dermis of most small cetaceans (porpoises and dolphins) is about 20 millimeters thick; in most large whales (including the fin and blue whales) the dermis is from 750 to 1,500 millimeters thick. The relative weight of the blubber in fin and blue whales is about 25 percent; in right whales it is 36–45 percent. The blubber of harbor porpoises ranges from 45 to 60 percent of the total body weight. Obviously the relative weight depends on the condition (fattiness) of the animal.

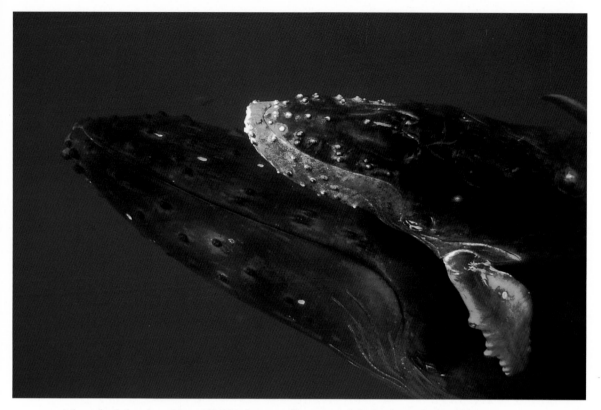

A humpback female and her calf. This is an excellent view of the knobs on the head that are characteristic of humpbacks. Each one of the knobs has a hair at its center. Some of the small white scars are probably the result of "cookie-cutter" shark predation. (Photo obtained under NMFS permit #987)

HOW LARGE ARE NEWBORN WHALES AND DOLPHINS?

Whales and dolphins are relatively large at birth, measuring a third to half of the length of the mother. They commonly double this length in the first year of life. The growth rate decreases thereafter until they reach physical maturity (at 15 to 20 years old). The largest whale, the blue whale, has the largest young. A blue whale is about 7.3 meters long at birth and weighs about 3 metric tons. It grows an average of 30 centimeters a week and gains weight at about 100 kilograms per day.

One of the smallest cetaceans, the franciscana, or La Plata river dolphin, is 70 to 75 centimeters in length and weighs 10 to 11 kilograms at birth.

WHAT DO NEWBORN WHALES AND DOLPHINS LOOK LIKE?

Newborn whales and dolphins look almost exactly like their parents. The head is relatively larger in newborn cetaceans (neonates), just as it is in other newborn mammals. Cetaceans have flaccid (limp, bent) dorsal fins and flukes when they are born. Observations of those in captivity tell us that these appendages become firm (turgid) in the first few hours, so if you see a dolphin or whale with flaccid appendages, you can be sure it is a neonate.

Another characteristic of newborn cetaceans is the presence of narrow transverse bands on the flanks, which have been called fetal folds. Why they are there remains to be explained satisfactorily. One conjecture is that the fetal folds result from skin on the inside of the curvature being folded while the dolphin or whale lies curled in the womb. However, the problem with that explanation is that the fetal folds should be on only one side of the calf, but they are on both sides. This would indicate that the fetus is capable of turning over inside the mother, which it may well be. If it turns over, though, you would expect the folds on the outside of the curvature to disappear. In any event the fetal folds appear to last from six weeks to a year and are not a reliable indication of neonatal condition.

HOW OLD DO WHALES GET?

Management biologists are concerned with the way that the life history of wild whales affects their capacity to reproduce under the pressures of being hunted. They are primarily concerned about the age at which whales become sexually mature and capable of reproduction. They age whales by estimating the growth layers in ear plugs of baleen whales or the growth-layer groups in the teeth of toothed whales (see *How Do We Estimate the Age of a Whale?*). In running the samples for determining the age of whales at sexual maturity, biologists have accumulated data that reveal both the age at which whales reach physical maturity and their maximum age. On the basis of individual recognition, tags, and biological data, researchers believe whales can live as long as 200 years.

Through individual recognition of killer whales in the Pacific Northwest, biologists have come up with longevity estimates of 50 to 60 years for males and 80 to 90 years for females. This radically extends the old longevity estimate based on tooth sections and makes more feasible the story of "Old Tom," an Australian killer whale who was said to be between 50 and 90 years old.

Stone harpoon points have been taken from bowhead whales that were killed in Eskimo hunts. At the time these whales were killed, stone points had not been in use for at least 75 years, indicating that the whale must have been even older than that. Recent studies using a biochemical technique (aspartic acid racemization) to age individual bowhead whales have indicated that some whales live in excess of 200 years.

The life span of most baleen whales is long. Fin whales have been recorded to live up to 94 years of age; sei whales, to 70; minke whales, to 60. Scientists expect the humpback has the potential for living much longer than the 48 years on record.

HOW DO WE ESTIMATE THE AGE OF A WHALE?

Toothed whales lay down one growth-layer group in their teeth per year. These layers are laid down in both the dentine and the cement. A tooth can be extracted from the animal, sectioned with a saw, decalcified, and stained so that the growth-layer groups can be counted to determine the age of the animal. However, there are some problems in determining the age this way; such a determination must always be treated as an estimate rather than an exact age, which can be known only for whales whose year of birth was directly observed. We can estimate the age of a whale lacking teeth by using a similar cross-sectioning operation on many other hard tissues, such as bone. The commonest tissue used is the ear bone (tympanic bulla). The disadvantage of using hard tissues is that they may represent only partial records, because in some species (humans, for instance) we know that growth in the early years involves resorption of some of the original bone.

In baleen whales, where teeth are not available, growth layers can be counted in the earplugs. Baleen whales secrete wax in their external ear canals, just as humans and other mammals do. Because the ear canal is closed, the wax remains and accumulates in layers. This forms the wax earplug. Researchers working with the earplugs of whales that had been taken by whalers hardened the plugs (fixed them in formalin), sectioned them, and read the growth-layer groups.

We can also estimate age by comparing the total length of an individual with a growth curve. This curve is produced by plotting the ages of known animals against their total length. This technique is usable only in the early years of an animal's life, when it is growing rapidly. In mammals the curve rapidly levels off as the animal approaches asymptotic, or maximum, length at physical maturity. In cetaceans it is useful for the first 10 years of life.

Cetaceans reach physical maturity when they cease growing in length. Growth continues as long as the vertebrae are growing. Vertebral growth occurs primarily along the suture between the centrum, or body, of the vertebrae and the vertebral epiphyses, or bony plates at the ends of the vertebrae. When the epiphyses fuse to the centrum, growth ceases and the animal is considered physically mature. Epiphyseal fusion starts at both ends of the vertebral column and finishes in the

TABLE 4. RELATIVE AGE STRUCTURE OF SELECTED WHALE SPECIES

Species	Sexual Maturity[a]	Longevity (years)	Body Length (meters)[b]
Fin	6–10	100	6–27
Blue	10	80–90	7–34
Humpback	5	96	4–18
Bowhead	?	211	3–20
Sperm	7–13	62	4–18
Gray	5–12	70	4.6–15
Bottlenose dolphin	7–14	46	1–4
Spotted dolphin	7–13	27+	0.8–2.6
Spinner dolphin	3–9	23	0.7–2.2

Note: Data were retrieved from dead specimens, except for the longevity datum on bottlenose dolphins, which is based on years in captivity.

[a] Age at first ovulation.

[b] Range includes length of newborn to maximum reliably reported length.

midthoracic vertebrae. If we examine this middle region of vertebrae in a cetacean skeleton, we can assess the degree of physical maturity that the animal attained. (See Table 4.)

HOW DOES THE LIFE HISTORY OF WHALES AND DOLPHINS COMPARE WITH THAT OF HUMANS?

On average, large whales have approximately the same life history as humans. They reach sexual maturity at about 8 to 10 years and physical maturity at about 20 years, and they have a longevity of about 50 to 100 years (see *When Are Whales Sexually Mature?* and *How Old Do Whales Get?*).

Our ability to determine the age of whales by the direct observation of known individuals is limited to their age at first reproduction, which is slightly different from age at sexual maturity. Age at first reproduction is 7.3 years for right whales and 4 to 6 years for humpbacks. Bottlenose dolphins have been observed to reproduce at 7 years. The data obtained from observations of live wild animals may differ from data obtained from dead animals.

HOW LONG DO WHALES AND DOLPHINS LIVE IN CAPTIVITY?

The life span (longevity) of whales and dolphins in captivity is generally shorter than that of cetaceans in the wild. The Marine Mammal Protection Act registers

TABLE 5. LONGEVITY OF SELECTED CETACEANS IN CAPTIVITY

Species	Age (years)
Bottlenose dolphin	46
Common dolphin	22
Pacific white-sided dolphin	21
Pilot whale	30
Killer whale	27

Source: R. R. Reeves and J. G. Mead, "Marine mammals in captivity," pp. 412–436 in J. R. Twiss and R. R. Reeves, eds., *Conservation and Management of Marine Mammals* (Washington, D.C.: Smithsonian Institution Press, 1999).

captive marine mammals and through the years has given data that have allowed the computation of meaningful longevity figures. In the past some oceanaria replaced animals that had died with other captives, but they passed the dead animals' names on to the new captives, which led to inflated longevity records. (See Table 5.)

Longevity in the wild is more difficult to determine. Long-term behavioral studies of such species as bottlenose dolphins and killer, humpback, and right whales are revealing a greater longevity than we had suspected (see *How Old Do Whales Get?*).

HOW WELL DO WHALES AND DOLPHINS SEE?

Cetaceans have excellent vision, both in water and in air. Their acuity (visual sharpness) is good, and they have binocular vision over at least part of the visual field. On the basis of the relatively small size of the eye, some authors in the past have stated that vision is reduced in cetaceans. Even though the eye of the blue whale is small relative to the size of the whale, in absolute terms it is still about four to five times as big as the human eye. Vision is not as useful to aquatic species as it is to terrestrial species because of the decrease in visibility under water, but it is still useful enough to warrant their retention of fully functioning eyes. The only cetaceans in which vision is reduced is the Ganges and Indus river dolphins (*Platanista*). The eye in these species has been reduced to an organ that merely senses light and darkness. The eye opening has been reduced to a slit 2 to 3 millimeters in diameter.

DO WHALES AND DOLPHINS DETECT COLOR?

Whales and dolphins do not have color vision (the capability of differentiating colors) and are most sensitive to light at the blue end of the spectrum.

A close-up of a humpback whale showing its eye. The white circular scars are left by barnacles. The grooves below the eye are the ventral grooves that are characteristic of all balaenopterids (rorquals and the humpback). They primarily serve to permit expansion of the mouth while eating. (Photo obtained under NMFS permit #987)

DO WHALES AND DOLPHINS SEE AS WELL IN THE AIR AS IN THE WATER?

If you, as a terrestrial organism, were to put your head under the water and try to see, you would find out that the optical properties of water are vastly different from those of air. In air, the cornea plays a very important role in bending the light rays so that they will converge on the retina. In water, the refraction of the cornea disappears, and the light rays converge well behind the retina. Cetaceans have overcome this, as can be seen in captive animals. Captive dolphins have excellent visual acuity both in the water and in the air, as is evidenced by their ability to follow the subtle visual cues of the trainer and catch objects that are thrown to them. How they do this is still a controversial topic. They may employ different parts of the eye that have different optical characteristics for seeing in water and in air, or they may have a unique double-slit pupil that responds to give excellent vision in both mediums.

A group of bottlenose dolphins swimming in shallow water. The dolphins arch their bodies to transfer energy to the flukes to swim.

WHAT COLOR ARE CETACEANS' EYES?

Cetaceans' eyes are brown. This appears to be the default state for mammals. Blue eyes in primates (humans) and certain dogs are an exception. Yellow eyes (in cats) are simply light brown.

HOW WELL DO WHALES AND DOLPHINS HEAR?

Whales and dolphins have extremely sensitive hearing. Hearing is a very important sense for cetaceans, because it enables them to perceive their environment over long distances, whereas vision in the water is commonly restricted to less than 100 feet. Cetaceans rely on the interpretation of sounds to provide information about their environment and its denizens. Studies of cetaceans in captivity indicate they obtain information acoustically that is comparable in detail to that which humans are able to obtain visually.

WHAT DO WHALES AND DOLPHINS HEAR?

Tests to determine the frequency range of dolphin hearing indicate that they hear very well over the same frequencies as humans (40 hertz to 18 kilohertz), and, in addition, they are sensitive to ultrasonic frequencies as high as the hearing range of bats (150 kilohertz). Comparable hearing data on whales are not available, but we assume that they can hear in the range of sounds that they produce. Some of the baleen whales produce very low and intense sounds but do not seem to produce the ultrasonic frequencies of toothed whales. Fin whales have been recorded producing calls with a frequency of 20 hertz, and blue whales with a frequency of 16.5 hertz.

HOW DO WHALES AND DOLPHINS HEAR IN WATER?

As cetaceans became adapted to a completely aquatic existence, anatomical changes began to take place in the ear resulting from differences in sound transmission in air and in water. Sound energy is reflected by any sudden change in the properties of the medium that it is penetrating. Sound in air can be channeled by a fleshy tube because of the difference in acoustic properties between the wall of the tube and the air. If you fill the tube with water, the acoustic properties of the water and those of the walls of the tube are very similar, so sound will pass through the walls easily. Because sound energy goes through tissue as readily as it goes through the water-filled ear canal, the cetacean canal became useless, and its supporting cartilage disappeared. The ear canal in whales and dolphins has evolved into a tiny thin tube. It is present in this form in all cetaceans.

The same situation holds for the external ear. In air it reflects sound, but in water it does not. With the loss of its value to direct sound, the external ear disappeared. Some seals have an external ear and some do not, but you have to remember that the seal ear must function in air as well as in water.

The role of the middle and inner ears of whales has changed drastically through time. The eardrum is no longer the primary mechanism for capturing sound. It has changed from a membrane under tension to a loose membrane that protrudes externally from the middle ear. This membrane looks remarkably like the finger of a rubber glove and is commonly called the glove finger.

The middle ear cavity expanded and became acoustically isolated from the rest of the skull bones by a combination of bony resorption and development of air sinuses. The exact mechanism of transfer of acoustic energy from the middle ear cavity and bulla to the cochlea is not well understood.

WHAT IS THE EFFECT OF UNDERWATER SOUNDS ON WHALES?

As human activities increase, so do the sounds associated with them. High-frequency sounds are rapidly attenuated in the water because of its transmission characteristics. As such, high-frequency sounds are only of limited trouble to whales. Low-frequency sounds, on the other hand, are less attenuated and may travel for tens or even hundreds of miles. One such source of low frequency sounds is ship propellers. The navy can, under optimal circumstances, recognize a vessel hundreds of miles away by its propeller noise. You can count on whales hearing the same sounds. There have been a number of proposals for stopping or suspending projects involving intense emissions of low-frequency sounds because of the possible effect on whales. This type of sound pollution can cause whales' social systems to change markedly and may affect their capacity to reproduce.

DO DOLPHINS USE THEIR JAWS IN HEARING?

In the mid-1960s a theory about the way sound is received by an odontocete gained acceptance. According to this theory, sound waves impinge upon the head of a dolphin, and the energy penetrates the soft tissues at the back of the lower jaw. Anatomically, sound penetrates to the jaw through a "window" of fatty tissue. The back end of the jaw is extremely thin and allows sound to pass through it into another area of fat and thence to the middle ear. This theory was confirmed with the mapping of areas of acoustic sensitivity of the head.

DO WHALES AND DOLPHINS TASTE THEIR FOOD?

Biologists long thought that whales and dolphins have a reduced sense of taste, because adult whales and dolphins lack taste buds entirely. But we know that taste is perceived by humans in a variety of situations that do not involve taste buds. In the 1970s, research in the Soviet Union demonstrated that the bottlenose dolphin is capable of differentiating sweet, salty, sour, and bitter tastes. Captive dolphins have preferences for different fishes, implying a more sophisticated sense of taste bordering on that of humans, who not only use stimuli from the taste buds and other receptors but integrate smell from the olfactory senses as well.

A Russian researcher recently demonstrated that the bottlenose dolphin has a highly effective sense, which he called quasi olfaction (*quasi* = seeming, apparent; *olfaction* = smelling). This sense operates through pits in the back and on the roots of the tongue and permits the dolphins to experience sensations that we would call smell, but it does not utilize the dolphin's nose.

DO WHALES AND DOLPHINS HAVE A SENSE OF SMELL?

The question of whether a sense of smell is present in whales and dolphins has been asked for a long time. When we deal with a special sense such as this one, it is best to define what we mean by "sense of smell." Smell is information about the chemical environment that is received through sensors in the nasal passage. The nasal passage in toothed whales has become heavily modified with the development of the structure involved in high-frequency sonar. The accessory air sacs in the nasal passage of toothed whales do not seem to have any function in chemical sensation, and it is likely that toothed whales are truly anosmic (lacking the sense of smell). Accordingly, in all toothed whales that we know of, the olfactory nerve has been completely eliminated.

Some baleen whales, on the other hand, have retained the part of their nasal passage that is concerned with smell (the olfactory area, the turbinal bones). The olfactory nerve is greatly reduced in baleen whales but is not completely absent (as far as we know). Baleen whales would have more use for smell than toothed whales, because they need to locate the areas of plankton concentrations, which affect the odor of the water. Of course, any whale could use the sense of olfaction only at the surface when it breathes, not when it is diving. (See *Do Whales and Dolphins Taste Their Food?*)

DO WHALES FEEL PAIN?

Pain is a highly subjective feeling. What one person feels as pain another might feel as only a relatively disagreeable sensation. Whales that have been stranded alive for several days and become severely sunburned have been noted to tremble in pain with only a gentle touch.

DO WHALES AND DOLPHINS HAVE A SENSE OF TOUCH?

The sense of touch is important for any animal, and whales and dolphins are more than adequately supplied with touch receptors in the skin. Touch is important in many behavioral contexts. Perhaps the most obvious with cetaceans is the role of touch in mother-calf interactions and mating.

DO WHALES HAVE A BUILT-IN COMPASS?

It has been demonstrated that some birds and fishes are sensitive to the earth's magnetic field and use that sensitivity to orient themselves. The extension of that line of research to other species has resulted in the question of whether cetaceans possess that sensitivity. If it is true, then disturbances in the magnetic field might be one of the causes for cetacean mass strandings. Unfortunately this is still a theory, and nobody has demonstrated that cetaceans are sensitive to geomagnetic stimuli. Magnetite has been found in some dolphin brains but not in others. Work has been done that attempts to demonstrate a correlation with stranding sites and geomagnetic disturbances, but the data are just not good enough to permit positive correlation much less demonstrate a causal relationship. It is an interesting theory, and it probably is at least partly true, but a lot of work still remains to be done in this field.

HOW DO WHALES AND DOLPHINS SWIM?

Whales and dolphins swim with vertical undulations of the spinal column, coupling this motion with the water by the broad horizontal expanse of their flukes. In contrast, most fishes swim with lateral undulations of the spinal column, which they couple with the water by using their vertically oriented caudal (tail) fin. The method among whales and dolphins is a direct result of their having evolved from a terrestrial mammalian ancestor that used vertical undulation in its locomotor pattern, which illustrates of one of the basic differences between the mammalian locomotor pattern and that of all other vertebrates. Primitive mammals tucked their legs under their bodies and progressed with vertical undulations of the spinal column instead of lateral undulations.

HOW DO WE DETERMINE WHAT WHALES ARE DOING UNDERWATER?

With the development of small radio transmitters to be used as tags to determine the location of a whale, monitoring the whale's physiology and transmitting data back to a receiving station became possible. Commonly tagged whales are equipped with a time-depth recorder (TDR). The output of the TDR permits researchers to determine how long a whale spends at the surface, how long it remains in the depths, and what its "dive profile" is; from this information researchers are able to theorize whether the whale was feeding or courting or traveling. A researcher can attach telemetry devices that monitor the heartbeat, water temperature and depth, speed of swimming, and a whole host of other parameters. The chief drawbacks are difficulties in fastening the tag to the whale, the expense of the tag itself, and the possibility that the attachment will modify the whale's behavior.

Work is currently being done with "crittercams," small camcorders that are attached to whales. The crittercam can record each part of a whale's environment as the whale experiences it. This device has obvious possibilities in investigations of feeding, courtship and mating, and reproductive behavior.

HOW FAST DO WHALES AND DOLPHINS SWIM?

Fin and blue whales can swim so fast that it takes a boat moving in excess of 16 knots (30 kilometers per hour) to catch them. Sonar records of fin whales indicate that they can sprint at 26 knots (48 kilometers per hour). Normally fin and blue whales can maintain 18 to 20 knots (33 to 37 kilometers per hour) for periods of up to 10 to 15 minutes.

Some whale species swim slowly, such as the right, humpback, and gray whales, which can seldom move faster than 5 knots (9 kilometers per hour). Sperm whales can cruise at 4 knots (7.5 kilometers per hour) and sprint in spurts at 20 knots (36 kilometers per hour). The champion is said to be the sei whale, which has been recorded at speeds of 35 knots (65 kilometers per hour) at the surface.

Common dolphins have been observed swimming alongside boats (not bow riding [see the glossary]) for a considerable period of time at 20 knots (36 kilometers per hour). That seems to be about the average for speeds of pelagic dolphins. Researchers Thomas Lang and Kenneth Norris trained Pacific bottlenose dolphins to run in an open-water environment, thus removing the spatial limitations of a pool but conserving the experimental controls. They found that the dolphins could sprint at 16.1 knots (29.9 kilometers per hour) in 7.5 seconds and could maintain a speed of 11.8 knots (21.9 kilometers per hour) for 50 seconds (Lang and Norris 1966).

Dolphins and porpoises, when swimming rapidly, launch their bodies out of the water in a low-angle leap in order to breathe. In this way they do not interrupt their swimming pattern with pauses to come to the surface. This is known as porpoising, which can be prolonged into spectacular series of leaps. Certain species of dolphins, in bouts of exuberance, not only leap but also rotate on their long axis in a manner that has become known as spinning. One such species, the spinner dolphin (*Stenella longirostris*), does this consistently enough to have been named for the behavior.

WHAT IS SOUNDING?

Sounding is the term used when whales and dolphins dive after having cruised at the surface. The animals have been "blowing" fairly regularly, and now they dive more deeply, usually showing their flukes as they do. Sounding is usually correlated with feeding, especially in the deeper-diving whales, such as sperm and beaked whales.

A humpback breaching. Breaching is the act of leaping out of the water and then falling back with a resounding splash. Breaching produces extreme amounts of noise, which advertises the presence of the whale. You can also see the enormous flippers that are characteristic of a humpback.

WHAT IS BREACHING?

Breaching occurs when whales and dolphins launch their bodies out of the water more or less vertically but with their backs angled slightly toward the water. They then descend onto their backs with a great crash. Breaching is a display function and probably serves as a method of primitive communication ("Hey! Here I am"). Breaches may occur singly or a number of times in quick succession. Breaching is to not be confused with leaping (where whales or dolphins reenter the water cleanly), with belly-flopping (where they enter the water right side up), or with "porpoising" (which facilitates breathing at high rates of speed).

Breaching also occurs in humpbacks as a part of their competitive display in courtship during the winter months. Large groups of males assemble to sort out hierarchical dominance, with breaching forming an active part of this process. Males are even known to breach in such a way as to fall backward onto one another.

HOW DEEP CAN WHALES AND DOLPHINS DIVE?

Most whales and dolphins spend much of their lives in the upper 100 meters of the ocean, which is the most productive zone in terms of food. However, the remain-

Left: Recent work has been done with time-depth recorders, which are fastened to a whale by suction-cup tags. Here we see a tag in flight toward its target, a bottlenose whale that has just blown. The tags are attached to a crossbow dart that is tipped with a suction cup. They are designed to come off after several hours and be picked up by researchers. *Right*: Some of the deep-diving whales, such as this sperm whale, have to swim vertically to get down to their accustomed depth of a kilometer or more.

ing huge expanse of ocean between 100 meters and 11,000 meters (the deepest point in the oceans) is bound to be occupied—and it is. It is the niche that sperm and beaked whales fill.

Sperm whales have been found suffocated after becoming entangled in submarine telephone cables at depths of 500 fathoms (1,000 meters). A study using acoustic transponders and sonar on sperm whales in the Caribbean Sea showed that most dives were to 400 to 600 meters, but at least one dive was to 1,185 meters, and another was possibly to 2,000 meters.

The only good record for depth of diving in beaked whales is the observation of dives up to 1,453 meters that was made on bottlenose whales off the coast of Nova Scotia.

A captive bottlenose dolphin was trained by the U.S. Navy to dive to various depths, return to the surface, and exhale its lung contents into a funnel for collection. The animal dove as deep as 535 meters during these tests. Bottlenose dolphins are not known as deep-diving animals, but these tests demonstrated that the animals can attain that depth.

Because whales and dolphins feed on aquatic animals, which tend to be active and occur at varying depths, the longer they can hold their breath and chase their prey, the better. Cetaceans have accomplished this in two ways. When they surface to breathe, they take in relatively more air than terrestrial mammals do. A human breath represents approximately a fifth of the lung capacity. Whales, on the other hand, utilize four-fifths of their lung capacity in a single breath.

One of the reasons cetaceans are able to dive as well as they do is that they have a greater amount of blood than terrestrial mammals of similar size (increased relative blood volume), and their hemoglobin carries more oxygen. Whales also store more oxygen in the myoglobin of their muscles than in the hemoglobin in their blood, the place that humans store oxygen. In addition, whales have an increased supply of myoglobin, an oxygen-binding pigment, which benefits their special circulation capacity and enables them to hold their breath longer than terrestrial mammals.

The muscles of whales can operate without oxygen and build up an "oxygen debt" for a longer period than the muscles of terrestrial animals can. Skeletal muscle can function for a time without oxygen (anaerobically), but cardiac muscle and the brain need a constant supply of oxygen. If they are deprived of free oxygen, the individual suffers either a stroke (a brain attack) or a heart attack. However, whales are able to shift the blood from the muscles to the brain and heart and utilize the oxygen that is stored in the hemoglobin to keep those two critical organs functioning.

Another anatomical peculiarity that is related to whales' diving capacity is the development of an extensive network of small veins and arteries in the chest (*rete mirabile*) (see *What Kind of Blood Vessels Do Whales Have?*). When a whale dives, the blood is shunted from the muscles into this network, where it circulates to the brain and vital organs.

DO WHALES AND DOLPHINS GET THE "BENDS"?

Whales are not subject to the "bends," because they do not breathe under increased pressure when under water. They breathe at the surface, then hold their breath while they dive (see *How Do Whales and Dolphins Breathe?*). Hence, they do not expose themselves to air at increased pressure and therefore avoid a buildup of excess nitrogen.

The bends, or caisson disease, is precipitated by breathing normal air (79 percent nitrogen and 21 percent oxygen) at increased pressure at depth. Increased pressure causes more nitrogen to be absorbed into the blood. When a person suddenly returns to normal atmospheric pressures with blood that is saturated with nitrogen, the nitrogen comes out of solution and forms bubbles. These bubbles tend to accumulate in the brain, central nervous tissue, and blood, causing severe pain and even death. Nitrogen bubbles can disrupt important neural pathways in the brain and spinal cord. The only cure is to repressurize the body and allow enough time for the bubbles to dissipate. Even with this treatment, the bends can cause permanent impairment.

A physiological experiment conducted on a bottlenose dolphin indicated that the dolphin's lungs completely collapsed at a depth of 100 meters. Therefore, below that depth, air in the respiratory system was excluded from the respiratory surface of the lungs, and no exchange of gases occurred between the blood and the air. This prevented the blood from absorbing an excess of nitrogen, which is what leads to the bends.

HOW DO WHALES AND DOLPHINS REMAIN BUOYANT?

Most whales and dolphins are positively buoyant at the surface. That is, with all other things being equal, whales and dolphins will float at the surface. When they dive, however, the air in their lungs compresses, and they become to a small degree negatively buoyant. They sink. Their buoyancy is held more or less constant by the store of blubber, which is slightly lighter than water. Blubber weighs on the order of 8–15 percent of the total body weight for a fin whale, up to 23 percent for a sperm whale, and up to 45 percent for a right whale.

WHAT IS CETACEANS' BODY TEMPERATURE AND HOW DO THEY MAINTAIN IT?

The normal body temperature (rectal temperature) in cetaceans varies from 36.4° to 37.2°C. In humans the normal range is from 36° to 38°C, with an average of 37.0°C. One way that cetaceans maintain their body temperature is through their insulating blanket of blubber (see *How Do Whales and Dolphins Remain Buoyant?*). Large whales have trouble getting rid of heat and frequently dump heat through the skin, mouth, and tongue. Small cetaceans, on the other hand, can have a problem with heat loss; this is offset by a countercurrent heat-exchange network, which reduces loss of heat from the skin.

The countercurrent exchange system of cetaceans is best seen in the dorsal fin. Arterial blood, destined for the skin of the dorsal fin, flows through the center of the fin and supplies the skin with oxygen and nutrients. In so doing, the blood cools as it travels, because the skin is cold. The cold blood returns via veins that sit next to the arteries, actually touching them. Heat from the warm blood in the arteries passes to the cold blood in the veins, thus warming the venous blood as it flows to the heart. At the same time, the arterial blood is cooled, resulting in a cool arterial blood supply to the skin. Thus the net exchange of temperatures is minimal. The countercurrent heat-exchange system also works for duck feet and human limbs. If any mammal wants to lose heat, it merely has to redirect the venous blood return away from the arterial supply.

HOW FAST DOES A WHALE'S HEART BEAT?

As animal species get larger, their heart rate decreases. This principle is illustrated with the following examples of heart rates: hamsters, 450 beats per minute; cats, 120 beats per minute; dogs, 120 beats per minute; humans, 72 beats per minute; cattle, 70 beats per minute; camels, 30 beats per minute; elephants, 35 beats per minute. Determining the pulse on a large whale is somewhat a problem. According to the only recorded account, a stranded fin whale's heart rate was 27 beats per minute, but that was determined as abnormally fast; the normal estimated heart rate was 8 to 10 beats per minute. The heart rate of a captive, 4,500-kilogram gray whale calf was 43 beats per minute.

Researchers have attempted to monitor the heart rate of a free-swimming whale. The last of these attempts occurred on an expedition that was funded by the National Geographic Society in 1956, when researchers tried to implant electrodes into a free-swimming gray whale with a crossbow, but they were unsuccessful.

HOW DO WHALES AND DOLPHINS BREATHE?

Whales and dolphins have lungs and breathe through their noses, like other mammals. The difference is that cetaceans spend most of the time submerged and can breathe only while their nostrils are above water. The entry to the nasal passage is the blowhole, homologous to our nostrils. The blowholes of cetaceans are normally closed when the animals relax, are submerged, are sleeping, and when they are dead. Opening the blowhole to breathe when the animal emerges requires muscular activity. The nasal plug must be withdrawn, and the nasal passages must be expanded. (See *How Do Whales' Lungs Differ from Our Own?*)

The commonest pattern of breathing and diving is exemplified by the fin whale. That species dives for 5 to 15 minutes then spends 5 to 10 minutes at the surface, breathing about twice a minute. It is not uncommon for sperm whales to dive for an hour or more and then spend about 10 minutes at the surface, breathing once every 10 seconds. The record for breath holding is held by bottlenose whales who

A blue whale at the surface, blowing. The breath is under pressure when it is expelled. It expands and cools, and the water vapor condenses as mist, making the blow visible.

were harpooned, dove for two hours, surfaced, and were "as fresh as if they had never been away" (Gray 1882: 727).

WHAT IS A BLOW?

The breath of whales is explosive and is known as the blow. The visible blow consists of three elements: (1) water vapor from the lungs condensing in the relatively cool air, (2) mucous particles accumulated from the walls of the air passages, and (3) seawater that remains on the whale after surfacing. The relative percentages of these elements vary according to the environment in which the whale finds itself and its level of activity. A human being exchanges roughly 10 percent of his or her lung capacity with every breath compared with 80–90 percent among whales.

CAN WHALE SPECIES BE IDENTIFIED BY THEIR BLOWS?

Sometimes species of whales can be identified by their blows. Whalers (and whale watchers) have long used the size, shape, and angle of the blow of whales to provide

The blowholes of some whales are slanted outward, so that the blow emerges as two plumes of mist, as in this humpback. You can see the difference in this and the preceding photograph of the blue whale blow.

preliminary identification. Often the blows extend 5 to 6 meters and are visible for several miles. An experienced observer can frequently estimate the height of the blow and get a good idea of the size of the individual. For instance, an adult blue whale 25 meters long generally has a blow that is 6 meters high. Under favorable circumstances, even dolphins can be located by their blows.

Most whales blow vertically; that is, the blow seems to hang perpendicular to the horizon. The only exceptions to this are the blows of some of the beaked whales and the blow of the sperm whale. The blows of beaked whales tend to be directed forward, and the blows of sperm whales are directed forward and to the left. The modifications in the sperm whale skulls and complex nasal passages of toothed whales result in a single external nostril and thus a single plume.

In contrast, all baleen whales have two external nostrils, creating a double blow. Despite this, the blow of most rorquals (blue and fin whales and their kin) merges into a single narrow plume shortly after it emerges, rising straight above the whale.

The humpback's double blow also often merges into a single plume. However, the double blows of the humpback, right, and gray whales, which are all baleen whales but not rorquals, tend to look short and bushy. Only in exceptionally calm weather do these blows appear differently. The low, bushy blows of right whales are generally double; that is, the two plumes, each coming from a single nostril, remain separate. The blows of gray whales are very similar to the blows of right whales.

HOW OFTEN DO CETACEANS BREATHE?

The breathing frequency of whales can be divided into three general classes: the deep divers, the shallow divers, and the puffers (see *How Do Whales and Dolphins Breathe?*). The deeper-diving whales, especially the sperm and beaked whales, remain submerged for an hour or two and then surface and pant, breathing regularly every 10 seconds for 5 to 10 minutes before diving again.

The second class of whales, the shallow divers, consists of most of the large baleen whales. These animals remain submerged for perhaps 5 to 15 minutes, then surface and breathe about once or twice a minute for 5 to 10 minutes.

The third type of breathing pattern is found in many dolphins. When they swim near the surface, they breathe about twice a minute. They can dive for up to 5 to 10 minutes and then "catch their breath" at a rate of about six blows per minute.

HOW DO WHALES KEEP FROM DROWNING?

Drowning occurs when an animal takes water into its lungs. Most people think that drowning results in lack of oxygen to the blood, but in fact, the dilution of the blood by water results in a decrease of blood pressure. That, in turn, results in decreased blood volume in the heart beat, causing unconsciousness. This is all brought about by inhaling water. All terrestrial mammals have a breathing reflex. That is, they can hold their breath until they pass out; then they breathe. Cetaceans are unusual in that they do not have a breathing reflex; breathing is a conscious activity that they must think about doing. If they do not, they become unconscious and die. But they do not drown in their unconsciousness; they suffocate.

HOW DO WHALES DIGEST THEIR FOOD?

Whales and dolphins, like the majority of mammals, have a multicompartmented stomach. The first compartment is just a dilation of the esophagus and is termed the forestomach. It is very similar to the forestomach of ungulates and serves as a reservoir to hold food. The second compartment, or main stomach, is the first of the true stomach compartments. The main stomach secretes acids and digestive enzymes and starts the digestion of food. The third compartment is actually two small compartments known as the connecting chambers, which are lined with a relatively simple epithelium. The connecting chambers serve as holding chambers to allow digestion to take place. The fourth compartment is known as the pyloric stomach, which is similar in anatomy to the pylorus in humans. It serves as another holding compartment to allow the digestive mixture to lose its acidity before it enters the first chamber of the small intestine, the duodenum. Further digestion and absorption of food take place in the small intestine. Here, food is in a fluid state. Water is resorbed into the body from the large intestine; then the remains are excreted. The feces of cetaceans are very fluid, unlike those of most terrestrial mammals. Most cetaceans do not have a pronounced difference between the small and large intestines. Some dolphins and baleen whales have a cecum, which marks the beginning of the large intestine, but other whales do not.

WHAT IS AMBERGRIS?

Ambergris is a substance that comes from the intestines of whales, and it is found only in the sperm whale. We do not know what role ambergris plays in the physiology of the whale. It has been suggested that it is a product of sick whales, but we have no proof of that. Because ambergris has the property of holding (fixing) smells, it has been extremely valuable to the perfume industry, often being sought after as a perfume fixative. The best-grade ambergris is a pale yellow substance that is found drifting on the ocean. Some pieces of ambergris are extremely large. The largest one known was a 421-kilogram mass that was taken from a 14.7-meter male sperm whale that was captured on 21 December 1953 at 58°23′ south latitude, 14°13′ west longitude in the South Atlantic Ocean east of the South Sandwich Islands. The ambergris that is taken directly from sperm whale intestines is crude, frequently smelling like well-seasoned manure and having numerous squid beaks embedded in it.

Ambergris was known for many years before its origin was determined. The name means "gray amber" and refers to the belief that it was a type of amber of uncertain origin. Originally, it was only found washed up on the seashore. The true nature of ambergris was discovered in 1725, when Paul Dudley wrote a letter from Massachusetts to England saying that he had discovered ambergris in the intestines of sperm whales.

Ambergris is a waxy substance that characteristically floats in cold water and melts in hot water well below the boiling point (63–66°C). It is insoluble in water but soluble in alcohol. Although ambergris can sometimes be recognized by these three characteristics, the best diagnostic is the "hot wire" test, which consists of the following steps: Heat a piece of wire or a needle in a gas or candle flame for about 15 seconds, then press it into the sample in question to a depth of about 3 millimeters. If the sample is genuine ambergris, a dark brown to black, opaque, resinous liquid will form around the wire and appear to boil. If the sample is wax, the liquid will be clear. Withdraw the wire and, before it cools, touch the wire with your finger. True ambergris will leave tacky, pitchlike strings sticking to the skin.

WHAT ARE THE DISEASES COMMON TO WHALES?

Whales are subject to most of the diseases that afflict all mammals. We are most familiar with the chronic diseases that form telltale scars, such as osteoarthritis. Whales also suffer from ulcers, tapeworms, skin diseases, kidney parasites, middle ear worms, pneumonia, peritonitis, mammary infection, and lungworm, just to name a few. Malignant tumors (cancer) rarely occur. When frequent tumors do occur in a population of whales, it is usually linked to pollution, as in the case with the beluga population in the Saint Lawrence River (Luoma 1989).

WHAT ARE THE PESTS AND PARASITES THAT INFEST CETACEANS?

Any species that is dependent on another species to maintain its existence, but does not harm the other species, is known as a commensal. Whales, as big as they are, provide a lot of space for "hitchhikers" that do not cause the whale any harm. One of the groups of animals that infest some types of whales are barnacles, which are crustaceans, like crabs and shrimps, but they attach to objects and form hard shells. Barnacles feed on particles that they catch with their feet as they sweep them through the water. In addition to attaching themselves to whales, barnacles also commonly attach to rocks in the surf zone or the sides of boats, where they can utilize the fast currents to increase the quantity of water that they sweep.

There are several genera and species of whale barnacles, some of which are related to the barnacles that infest sea turtles. Some of these occur only on a particular species of whale, such as *Cryptolepas* on the gray whale; some are peculiar to whales only, such as the acorn barnacle *Coronula* and the pseudostalked barnacle *Xenobalanus*; and some, such as the stalked barnacle *Conchoderma*, are cosmopolitan in their choice of attachments, being found on ships and offshore buoys as well as on whales. It is only the acorn barnacles that can attach directly to the skin of whales because of the position of their hard shell, which is at the base, where the barnacle attaches to the whale. In stalked barnacles the shell is in the body, above the stalk, so they must have a hard substance on which to attach, such as the teeth of whales or the shells of other acorn barnacles.

One of the most common external commensals, of both whales and dolphins, is whale lice (cyamids). These small animals, mostly less than a centimeter long, are crustaceans that have become adapted to live on the skin of cetaceans. They presumably feed on sloughed-off skin and diatoms. Whale lice are extremely common on right and gray whales but occur only sporadically on many other species. In many cases any given species of cyamid occurs on only one species of whale.

The white areas in this picture illustrate the whale lice that live on the callosities of this right whale's head. The callosity at the far left of this picture is called the bonnet and lies on the tip of the upper jaw. The actual tissue of the callosity is seen as small bits of dark gray matter that projects through the blanket of whale lice.

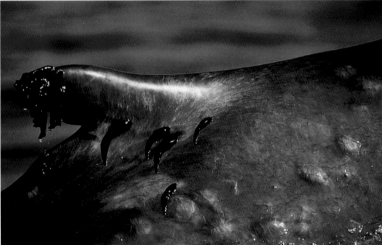

The dorsal fin of a blue whale off Mexico, showing the unusual occurrence of two types of commensal organism. The black animals on the tip are pseudo-stalked barnacles, which are about 5 centimeters long. You can see the small white base of the barnacles where they attach to the whale. The other small black animals lower on the flanks of the whale are remoras, fish that are also known as shark suckers. The barnacles are permanently attached; the remoras are attached by suction and move around on the whale to feed.

A large remora attached to the belly of a young spotted dolphin. These fish are "hitchhikers" and do no harm to the dolphin.

Another commensal organism that infests whales and dolphins is fish commonly known as remoras, or shark suckers. These fish attach to their hosts by means of a dorsal fin that has been modified into a suction device. Remoras are only along for the ride and frequently detach themselves from their host to feed.

Whales, like most animals, suffer from a variety of internal parasites, some of which can be serious. We have found that virtually all of the coastal bottlenose dolphin populations suffer from a pancreatic parasite that eventually destroys the pancreatic tissue. Although we cannot know how the dolphin experiences this condition, we do know that inflammations of the pancreas are exceedingly painful and debilitating in man.

Some parasites are relatively benign, such as the roundworm, *Stenurus*, that inhabits the middle ears of Atlantic white-sided dolphins, harbor porpoises, and pilot whales. These organisms were discovered in necropsies of mass-stranded Atlantic white-sided dolphins in the 1970s. Since proper functioning of the ear is crucial to echolocation, and hence feeding and navigation, scientists (and the news media) thought they had discovered one of the causes of mass strandings. However, this conjecture became less certain when it was later learned that the Russians had published on similar infestations of the same worm in normal fishery-taken dolphins.

Penella, a type of copepod (a group of small shrimplike crustaceans), has become an external parasite on fishes and whales. Larva of *Penella* burrow into the whale's skin and develop a single appendage that protrudes from the skin like a stalk of wheat, which consists of gills and reproductive organs. Also of interest is another species of *Penella* that burrows into the pupil of Greenland sharks (*Somniosus microcephalus*), and its appendage protrudes from the eye of the shark.

Some parasites are less benign, such as the roundworms that infest the kidneys of beaked whales and rorquals, causing blockage and infection in the vessels of those organs and an appreciable loss of function.

DO WHALES CRY?

It is frequently stated that cetaceans do not cry, because they have no lacrimal, or tear, glands. Besides that, they are in the water all the time, so do not have a problem with their eyes drying out. In fact, whales' eyes are provided with large Harderian glands (different from lacrimal glands), which secrete abundant viscous fluid. We have seen a live stranded humpback whose eyes were running over with secretions from the Harderian gland—secretions that we equate with tears. The "tears" of whales and dolphins are thick, which prevents them from washing out.

A beluga that has accidentally stranded in the Canadian Arctic, shown "crying." The beluga's eyes are irritated. The secretions from the eyes of whales and dolphins are extremely thick and appear as streams of mucus. The ear of this whale is seen as a dark hole, like a mark made by a pencil to the right of the eye.

HOW SMART ARE WHALES?

Whales have the largest brains known. The human brain weighs about 1,200 grams. The smallest cetacean brain appears to belong to the franciscana, or La Plata river dolphin. Researchers weighed 14 brains of these dolphins and found their average weight to be 221 grams. Through further research on other cetacean brains, the following weights were determined: 1 harbor porpoise brain, 500 grams; the average weight of 20 bottlenose dolphin brains, 1,824 grams; the average weight of 3 narwhal brains, 2,997 grams; the average weight of 2 killer whale brains, 5,059 grams. The largest whale brain, that of the sperm whale, has been recorded at a maximum of 9.2 kilograms, with 7.9 kilograms as the average weight of 20 sperm whale brains.

The anatomy of the whale brain is very similar to our own, but how the brain functions and whether its size is related to intelligence are the subjects of debate. We know that, as an animal increases in size, its organs also tend to increase, particularly the organs that deal with respiration, digestion, excretion, and other physiological functions. The brain, on the other hand, need not increase in size just to handle the energy requirements necessary for maintaining the body. One method used to describe the possible link between brain size and intelligence is to compare the weight of the brain with the weight of the body and produce a brain–body weight ratio. For example, the brain–body weight ratio in man is about 0.0171 to 1 (average body weight, 70 kilograms; average brain weight, 1.2 kilograms); the ratio in a bottlenose dolphin is about 0.0087 (average body weight, 210 kilograms; average brain weight, 1.824 kilograms); and in a sperm whale the ratio is about 0.000188 (average length, 15 meters [Berzin 1972: 30]; weight at that length, 42,100 kilograms; average brain weight, 7.9 kilograms). By this measure, man has a larger brain–body weight ratio than a sperm whale. But many scientists feel that the brain–body weight ratio is too simplistic to be of much use. Just because an animal has four times the muscle weight does not necessarily mean that it uses four times the brain weight to control its muscles. About one-fifth of a whale's weight is blubber. Does that blubber account for much brain activity?

The increased brain size of cetaceans may be the result of their ability to use a sophisticated auditory system. Toothed whales use an active echolocation system (sonar), whereas baleen whales appear to use a passive system. Studies indicate that the amount of information that the cetacean derives acoustically from its environment is comparable to the amount that humans derive through vision. Yet vision is an extremely simple system, possible to duplicate electronically with tiny video cameras. Duplicating the acoustic sense of whales would take many rooms full of electronics. Whales' ability to interpret their sonar information, rather than increased intelligence, may account for their larger brain size.

Humans have encountered profound problems in trying to measure "intelligence" among individuals of their own species and have had less than reliable success (for example, IQ measurements). When we hardly understand our own intelligence, we would be extremely arrogant and suspect to make any speculations about the relative intelligence of other species.

HOW DO WHALES REPRODUCE?

The reproductive organs of a female whale are similar to those of other mammals in many respects. They consist of two ovaries where eggs are produced and two uterine (Fallopian) tubes that lead away from the ovaries into the uterus, which narrows into a neck called the cervix. The cervix opens into a vagina lined with a series of folds. Within the ovary, egg cells (oocytes), encircled by other kinds of cells, form follicles that enlarge. The ovary becomes firmer and granular on its outer surface as the follicles become larger and mature. One of these greatly enlarged follicles, at a propitious time in the breeding season, will burst and release a single oocyte to be fertilized, not multiple oocytes as in dogs and cats. This is known as ovulation, and we think it occurs only once a year.

The egg travels to the end of the uterine tube. Meanwhile, the follicle forms a large gland, called the corpus luteum (literally, yellow body), which secretes hormones that prepare the uterus to receive the egg for implantation if it is fertilized. Whether or not conception occurs, the corpus luteum eventually degenerates, producing a scar on the ovary called a corpus albicans (literally, white body), which appears wartlike on the surface of the ovary. This is normal ovarian activity in any mammal. Whales and dolphins are unusual because the scars, corpora albicantia, are retained throughout the life of a female; therefore, by counting the scars, we can estimate how often the whale ovulated. However, these data still fall short of determining whether the animal ever actually became pregnant or was even capable of reproducing; just because a cetacean is capable of ovulating does not mean the female is capable of reproducing (see *When Are Whales Sexually Mature?*).

The reproductive cycle of cetaceans is divided into two phases: a resting phase, during which reproductive activity ceases, and a phase of sexual activity, which includes courtship and mating followed by gestation (pregnancy), birth, and lactation (production of milk). When the calf is weaned and lactation ceases, the animal returns to its resting stage. The gestation period of most cetaceans is about a year, followed by a period of lactation of at least six months. Although some cetaceans are capable of conceiving immediately after they give birth, and thus are simultaneously pregnant and lactating, most are not. Mating is seasonal, occurring once a year during the winter of whichever hemisphere the whale occupies. This means the average time between calves is two years (one year of pregnancy, six months of lactation, six months resting).

Of course, reproduction in whales (or any species) does not begin with mating. The act of copulation does not mean that a whale is sexually mature. Many immature whales will attempt to copulate before they are physically or physiologically or biologically capable of bearing young. The maturation process takes time. A female cetacean usually starts producing eggs in the left ovary. After a number of ovulations, the right ovary becomes mature. Following maturation of the right ovary, the ovaries more or less alternate the production of eggs.

Statisticians use reproductive data to determine the minimum age at first ovulation, which represents the earliest age at which it might be possible for the female to become pregnant, and the mean age of females at sexual maturity, which represents the age at which 50 percent of the animals are sexually mature.

In males, the reproductive organs, the testes, develop gradually; the sperm form gradually; growth is prolonged and seasonal. Because obtaining data on males is more difficult, research on male reproduction issues is meager. Also, we have less reason to be concerned about or interested in the maturation of males, because fewer males are necessary for the maintenance of the species. In terms of population dynamics, there is no need for large numbers of males; one male can easily mate with many females.

WHEN ARE WHALES SEXUALLY MATURE?

To understand when whales are sexually mature, it is necessary to understand what is meant by *sexual maturity*, but opinions differ about the meaning of this term. Some scientists (for example, those in stock management of whales or dolphins) define sexual maturity as the first ovulation, or production of eggs. Telltale scars found on the female ovary are signs that the whale has ovulated (see *How Do Whales Reproduce?*). Each time a whale ovulates, a scar is produced, and it is the presence of these scars that cetacean biologists look for when investigating accidental catches of whales in gill nets or in tuna purse-seining activities or when taking data on whales that strand. Much is known about the reproduction of cetaceans, because numerous animals have been examined after death, providing data on their reproductive organs, so we also know that even though a whale may be capable of ovulation and shows the related scar, she is not necessarily capable of reproducing. She may not have copulated; her eggs may not have been fertilized; and/or the rest of her body may not be mature enough to sustain a pregnancy. Therefore, other scientists believe that sexual maturity is attained when a cetacean is capable of producing viable offspring. The ability of a cetacean to produce offspring could be called reproductive maturity.

DO WHALES COURT?

The best-known instance of courtship behavior is the singing of humpbacks. Solitary males sing elaborate courtship songs. Male humpbacks engage in other kinds of behavior that presumably fills a role in courtship, such as stroking with their flippers. Courtship repertoires for cetaceans may also include chasing, rubbing, slapping the water with fluke or fins, breaching, butting heads, rolling, lunging, or even having an erection. Some of this behavior may serve to threaten other males, and some may serve to attract females.

Attention has been given to the casual sexuality with which male cetaceans seem to greet the world. The conjecture is that a male whale may use his erect sexual organ to manipulate another animal, which seems reasonable given his lack of hands and feet to use for this purpose. This manipulation may be a socially recognized greeting; it may be play; or it may be investigative behavior.

Although such sexual behavior is much more difficult to observe in females because of the lack of an external "signaling" member (a penis), female dolphins in captivity have been observed to present their genital area openly to other individuals, male or female. Since this area is one of the most tactilely sensitive areas of the body, we should not find such behavior surprising.

Whales and dolphins have behavior we would classify as courtship. The limitation of our understanding lies in our inability to recognize it as such.

HOW DOES MATING OCCUR?

All whales and dolphins mate underwater lying belly to belly. They may copulate while swimming, or they may be relatively stationary. The amount of time they remain coupled varies. It may be extremely brief (a matter of seconds) and is limited by the necessity to surface in order to breathe.

From an extremely young age, whales and dolphins engage in many copulations that do not result in pregnancy. Even nursing calves are frequently observed to copulate with their mothers. Cetaceans seem to use copulation as part of socialization rather than solely for reproductive purposes.

Then there is the question of making the distinction between courtship and mating behavior. If none of the early copulations of right, gray, and humpback whales

Bottlenose dolphins mating in captivity. This posture is typical of cetacean copulation.

result in conception, then, according to biological definition, that behavior would be courtship rather than mating behavior.

DO WHALES EVER HAVE TWINS?

Whalers have recorded instances of twin or triplet fetuses from females examined after they were taken, but we have no actual account of twin calves. Because of the relatively large size of a whale calf at birth, the chances of a successful delivery of twins are relatively remote. The presence of multiple fetuses probably indicates that all but one of the fetuses would have to abort or the cow would die as well. The report of twin humpback calves from Scammon's account of whales off the northwest coast of North America is fabulous (in the sense of occurring only in stories or fables). Scammon, in discussing the position of the cow whale when nursing, says: "In this way two calves *would* be enabled to obtain their nourishment at the same time" (Scammon 1874: 45). He goes on to illustrate that premise in plate 9 (facing page 48). It is clear that Scammon thought the situation could exist and had his artist draw it that way.

The incidence of twin fetuses in blue and fin whales is said to be about 1 in 100 pregnancies; triplets, 1 in 1,000; quadruplets, 1 in 10,000; and one case of quintuplet fetuses has been observed. The incidence of twinning in humans is about 1 in 80.

An account exists of a 22.5-meter female blue whale with six fetuses: two males of 3.5 and 2.39 meters, and four females of 3.15, 2.9, 2.82, and 2.21 meters. She was taken in the Antarctic on 21 February 1953, and the medical officer of the expedition was present and confirmed the data (Vangstein 1953). There is also a report of a fin whale with six fetuses taken in Hellisfjord, Iceland, on 10 July 1909 (Collett 1911–12).

ARE NEWBORN WHALES RELATIVELY LARGER THAN OTHER NEWBORN MAMMALS?

Whales have relatively small calves, even though the absolute size of the calf is large. Large animals are much more easily measured by length than by weight; the opposite is true for smaller species. For this reason, we know the relative size of newborn cetaceans by their length (28–60 percent of the mother's length) and the relative size of smaller mammals by their weight: for example, pigs weigh 2.5 percent of their mother's weight; cows, 4.6–6 percent; humans, 5.5 percent; sheep, 8 percent; and guinea pigs, 12 percent.

Unfortunately, any kind of meaningful comparison of the sizes of newborn cetaceans and other newborn animals requires some mathematical computations. As a general rule, weight is proportional to the cube of length; so the relative birth weight should be proportional to the cube root of the relative birth length. The cube roots of 28–60 percent, pertaining to the newborn whale, are 3–4 percent, which fall on the low end of the relative weights of terrestrial animals, as we can see from the examples shown above.

HOW LONG IS THE GESTATION PERIOD OF CETACEANS?

The gestation, or pregnancy, of whales and dolphins is about one year. Where data exist, they vary from a low of 8 or 9 months (in *Platanista gangetica,* Ganges river dolphin) to a high of about 16 months (in *Berardius bairdii,* Baird's beaked whale). More studies on this subject are needed to answer questions such as whether cetaceans use delayed implantation, a mechanism where a female ovulates and conception takes place, but development is halted and the embryo does not implant in the uterus, thus providing a way to prolong the gestation period to allow it to synchronize with climatic seasons.

HOW DO MALE AND FEMALE CETACEANS RECOGNIZE ONE ANOTHER?

Whales, like other mammals, can recognize each other as male or female by their different behavior. Many subtle behavioral characteristics of each sex contribute to sexual recognition. Because the range of vision is limited underwater, cetaceans depend on their hearing for recognizing the different sexes. Of course, some sexual dimorphism in external features does exist. For example, male killer whales have tall dorsal fins, almost 2 meters tall in some individuals. Also, there is no difficulty in recognizing male sperm whales; the male is 40 percent longer than the female. Narwhals have no physical recognition problems either, because the males sport a tusk that is nearly half as long as their bodies, and females normally do not have tusks. We do not really understand the meaning of sexually dimorphic features in cetaceans, such as the bulbous shape of the head in pilot whales or the presence of a postanal keel in males of many dolphins.

Probably the principal way that cetaceans recognize both the presence and the sex of other animals is through acoustics. It has long been recognized that dolphins within a group have distinctive vocalizations, called signature whistles. Individual dolphins can recognize one another by their vocal characteristics, which also carry sexual information. Cetaceans in general use vocalizations to advertise their sex and sexual condition, just as birds do. The best-known examples of this are the songs of the humpback whale (see *Why Do Some Whales Sing?*).

Two adult male killer whales flanking females or juveniles. The adult males have taller dorsal fins than females and juvenile males.

Five male narwhals swimming abreast. Female narwhals can easily recognize males by their tusks.

HOW DO WHALES AND DOLPHINS CARE FOR THEIR YOUNG?

The mother-calf tie is extremely well developed in whales and dolphins and appears to last beyond the weaning stage in some species. Most cetaceans are highly social animals, and the calf becomes an important part of its behavioral group. The role of other group members in the birth of a whale or dolphin has been documented. Dolphins that are known colloquially as midwives or aunties often assist the newborn calf in getting its first breath. The role that these assistants play in further development of the young may be substantial but is not yet fully documented scientifically.

HOW DOES LACTATION TAKE PLACE IN CETACEANS?

Whales and dolphins normally suckle for around six months. This is the "obligatory" suckling period, or the period that calves depend only on the mother for milk. At the end of the obligatory period, some calves continue to take milk on an opportunistic basis. When calves begin to take solid food, their stomach contents are frequently a mixture of remains of solid food and milk. This represents the beginning of the weaning period. Some species can extend their lactation period, such as the short-finned pilot whale. A study that was done on a Japanese population of short-finned pilot whales revealed that the obligatory lactation period lasted from 6 months to 1 year, but some calves continued to suckle for up to 10 years (Kasuya and Marsh 1984).

Lactation in some species takes place in special nursery areas. Notably, most gray whales calve and nurse in the sheltered lagoons of Baja California. Other species have nursery areas elsewhere, such as the southern right whale near the Valdés Peninsula of Argentina. Migratory species such as the fin and blue whales tend to calve on their wintering grounds and continue lactation during migration to their summer feeding grounds. Females with calves usually leave the wintering grounds later than females without calves.

Individual bouts of lactation take place when the cow presents the area that contains the nipples (the area around the genital slit) to the calf. The calf opens its mouth, and the mother squirts the milk into the calf's mouth. The exact role of the calf in initiating a bout of lactation is poorly understood. It is probable that the cow needs some tactile stimulation to begin the flow of milk, but we know that the calf does not "suck." The lips of the calf are stiff and cannot form a seal around the nipple; hence, they cannot form the suction necessary to extract the milk.

The female has a prominent muscle encircling the mammary gland. This muscle

A humpback "spy hopping" near a whale-watching boat off the coast of southeastern Alaska. Because of the lack of splashes around the humpback, you can tell that he has held this position for some time and is not in the act of breaching.

is perhaps similar to the supramammary muscles of dogs, which may function in helping the milk to flow. In cetaceans the muscle is very large, and the female whale or dolphin actively ejects the milk. Cetacean milk is high in fat (about 40 percent), which may keep the milk jet from dispersing in the water.

WHY DO DOLPHINS AND PORPOISES SEEM TO SMILE?

One of the aspects that has endeared porpoises and dolphins to the public is that they seem to be always "smiling." We instinctively react favorably to a facial expression in which the lines of the mouth are turned up (a "happy" face) as opposed to one in which the lines are turned down (a "sad" or "frowning" face). As primates we rely on facial expressions of other primates, a nonverbal communication, to supplement our understanding and enhance our lives.

Porpoises and dolphins, on the other hand, do not depend on facial expressions for communication, primarily because there is such limited visibility in water. As a consequence dolphins and porpoises have lost the musculature that would permit changes in facial expression and instead have developed thick and immobile skin that cannot change. Dolphins have a "plastic smile" no matter what is on their mind.

A killer whale "lobtailing," or beating the water with its tail. This activity does not result in the forward motion of the whale and is commonly repetitive.

WHAT IS SPY HOPPING?

Spy hopping is a term used to describe the behavior of whales and dolphins when they stick their head out of the water, pause, and look around. The term was first used by Yankee whalers in referring to that activity. The fact that whales do spy hop is one of the earliest and most convincing arguments that their vision is useful both above and below the water.

WHAT IS LOBTAILING?

Whales and dolphins sometimes smash the surface of the water with their flukes in an act that has become known as lobtailing. It appears to be a form of communication that the whale uses to identify its presence. The impact of the flukes of a large whale can be heard for miles. Lobtailing has also been observed as an aggressive display under certain circumstances.

WHAT DO WHALES EAT?

Whales and dolphins are carnivorous in that they consume only other animals. Some of these animals may be extremely small, but they are animals nonetheless.

Whales and dolphins prey on fishes, squids, and crustaceans (shrimp, crabs, and their relatives). The size range of their prey is enormous. Killer whales have been known to feed on blue fin tuna (up to 3 meters in length), and sperm whales on giant squid (with a mantle length of up to 5 meters). The larger toothed whales also consume birds and marine mammals, such as dolphins and seals and even other whales. The prey size ranges downward to small schooling fishes like anchovies, small squids, and many crustaceans on the order of what we would call plankton.

We have attempted to present the normal digestible diet of cetaceans. Some whales and dolphins appear to sample anything they can swallow or to swallow inedible material accidentally while pursuing edible prey.

Whales sometimes scavenge in ships' garbage. One of the authors, while working for the Fisheries Research Board of Canada at a whaling station in Newfoundland, confirmed this behavior. On 3 May 1972, one of the small whaling vessels brought in a young female minke whale that had numerous scars on its palate. Upon necropsy, a piece of steak bone was found embedded in the palate. It appeared that the scars were the result of this whale's having eaten other sharp pieces of garbage.

A century ago some biologists expanded the definition of the order Cetacea to include a group that was known as the herbivorous cetaceans, a historical variation that sometimes confuses people. The herbivorous cetaceans, which we now call Sirenia (sea cows, manatees, and dugongs), are no longer believed to be related to the true cetaceans. There is, however, good evidence that gray whales may ingest seaweed, and there are anecdotal tales about the Amazon river dolphins (*Inia geoffrensis*) occasionally eating fruit.

An unusual feeding behavior has been seen in bottlenose dolphins, where a small group of them will herd fish toward a smooth mud bank. The dolphins chase the fish and, in doing so, generate a wave that carries the fish onto the bank. Then, when the wave recedes, the fish are left stranded, and the dolphins snap them up.

The stories of killer whales preying on wounded whales and eating their tongues are common among whalers and whaling authors. We have yet to see any first-hand account of this behavior. A well-known biologist circulated the story that a killer whale was found to have "no less than thirteen complete porpoises and fourteen seals in the first chamber of its stomach (6½ feet × 4¾ feet). A fifteenth seal was found in the animal's throat" (Slijper 1962: 274–275). He emphasized the story with an illustration of a killer whale with the 13 porpoises and 14 seals drawn to scale. It turns out that this story was based on the findings of a Danish biologist who had found "remains of" 13 porpoises and 14 seals in a killer whale's stomach. Some of these remains consisted of bones that had been in the stomach for some time.

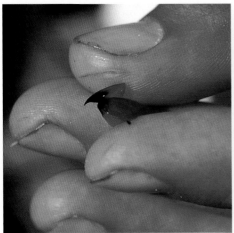

Left: The upper beak of a squid. Cephalopods make up a significant portion of the biomass of the oceans and are preyed upon by most large predators. Scientists are capable of identifying the types of squid that were eaten by a whale by examining the squid beaks in its stomach. *Right:* Bottlenose dolphins in salt marshes and rivers in the southern United States cooperatively drive small fishes toward the banks. Then they rush the fishes, generating a wave to propel them out of the water and onto the bank. There, they can pick up the fish with their beaks and eat them.

Killer whales have been documented on film pursuing seals onto the beach and then catching them and tossing them into the air (see *How Do Cetaceans Catch and Eat Their Prey?*). It has been known that certain populations of killer whales eat seals and sea otters and, to a lesser extent, porpoises and dolphins. Their predation upon larger whales is thought to be opportunistic and usually involves large whales that are injured or in some way compromised. There are scattered reports of killer whales preying on the young of large whales, even a blue whale.

In October of 1997 a research vessel of the National Oceanic and Atmospheric Administration happened upon a group of nine adult female sperm whales that were gathered in a "rosette." Whalers had often talked of this pattern of sperm whales, which consists of whales gathered in a tight circle and facing inward when confronted with danger. The danger in this case happened to be 40 to 50 killer whales, who systematically attacked and killed the sperm whales, eating some. This was observed by the scientific crew of the vessel, consisting of a party of cetologists who had been looking for sperm whales (Pitman and Chivers 1999).

Recently there has been a tendency to substitute the vernacular name "orca" for the killer whale. The feeling behind this is that *killer* carries certain derogatory connotations and is inapplicable to *Orcinus orca.*

A smaller relative of the killer whale, the false killer whale (*Pseudorca crassidens*), has also been known to prey upon other cetaceans. In 1972 whale-watching boats observed four or five false killer whales eating a live humpback whale calf off the coast of Hawaii.

Although whales are not generally known to prey on exceedingly large sharks, such as the basking shark (10 to 15 meters in length) or the whale shark (10 to 18 meters), there are some exceptions. A team of biologists reportedly encountered sperm whales feeding on a megamouth shark (*Megachasma pelagios*), which is a recently discovered shark species (1976) that attains a length of about 5 meters, has small teeth, and feeds on plankton (Pecchioni and Peoldi 1998). We also have a controversial report of a recent (1997) encounter between a white shark and killer whales near the Farallon Islands, California. That incident appeared to be motivated more by the killer whales' aggression toward the white shark than by their opportunity to feed on its carcass (Grubs 1997).

Some baleen whales tend to feed on crustaceans, usually known as krill, which range in size from 1 to 5 centimeters. The largest whale, the blue whale, feeds chiefly on *Euphausia superba* in the Antarctic, one of the larger species of krill, measuring 5 centimeters. Blue whales frequently eat other prey but concentrate on species that are on the order of 5 centimeters long. Right whales and sei whales eat the smallest prey, copepod crustaceans. Copepods are 1 centimeter or less in length, about the size of a grain of rice. Baleen whales also feed commonly on schooling fishes, such as herring, and do occasionally feed on small squids.

WHAT ARE KRILL AND BRIT?

Krill is a Norwegian term that has been applied by whalers to small crustaceans (euphausiids) of the family Euphausiidae. Euphausiids are shrimplike in form and range up to 6 centimeters in size. They concentrate by the millions in areas of upwelling and constitute an enormous food source.

Brit is an obscure term, of English origin, that was used to refer to "whales' food." A naturalist whaler, Charles Melville Scammon, spoke of brit in the following words in his discussion on bowhead whales: "*animalculae* (termed by the whalemen) 'right whale feed,' or 'brit'" (Scammon 1874: 54). The term was in common usage among whalers in the early nineteenth century. It appears to have come from English whalers who discovered "sea snails" (pteropods, minute mollusks with spiral shells) in the water and stomach of the bowhead whales. Another whaling naturalist, William Scoresby, mentioned the pteropod genus *Clio* in his work on the Arctic seas but did not mention the term *brit*. *Brit* was also used to refer to larvae and juveniles of several fishes.

Some of the shrimplike organisms that are present in large schools where upwelling brings nutrients to the surface are euphausiids, or krill. These animals are only a few centimeters long, but in millions they form the food base for some baleen whales.

HOW DO CETACEANS CATCH AND EAT THEIR PREY?

The method of capturing and eating prey differs between the two suborders of cetaceans, Odontoceti and Mysticeti. Odontocetes (toothed whales and dolphins) capture their prey individually, whereas mysticetes (baleen whales) capture their prey as schools.

Lunge feeding is a technique used by baleen whales whereby the whale approaches a school of prey, either fish or krill, and thrusts its body forcefully into the school with its mouth open. Then it closes its mouth and swallows the prey. This type of feeding is commonly used by fin whales and humpbacks. This is an extension of the feeding technique known as swallowing, where the whale detects a concentration of prey, opens its mouth, takes a mouthful of prey and water, closes its mouth, and expels the water through its baleen.

Skim feeding is another technique used by baleen whales. The whale swims along with its mouth open for a number of minutes in an area where prey is present but not in sufficient concentrations to make lunge feeding practical. This type of feeding is used by sei and right whales to feed on copepods.

Bubble netting, which has also been described in humpbacks, is a technique in which the whale swims in a large circle blowing bubbles to form a net, or wall, of

The most impressive act of cooperative feeding behavior is the lunge feeding of humpbacks, where a group of humpbacks charge vertically upward through the water into a school of prey that they have herded into a bunch at the surface. Then the humpbacks erupt from the water with their mouths full of fish.

Two fin whales feeding. Another way of feeding involves swallowing, which consists of an individual whale detecting a school of prey, opening its mouth, and engulfing the prey. The fin whale on the right has turned over onto its side and just swallowed a mouthful of prey; the one on the left did the same thing moments earlier and has now turned right side up and is preparing to blow.

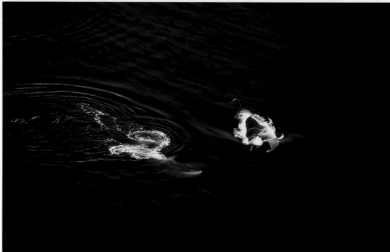

bubbles that serve to concentrate the prey, who are afraid of the bubbles. Having done this, the whale dives and come up through the center of the bubble net with its mouth open and engulfs the prey. Sometimes a group of whales will work cooperatively using this technique.

Toothed whales and dolphins, of course, use different feeding techniques, although they, too, almost always consume their prey whole, because their teeth are unsuitable for chewing. Rather, their teeth enable them to capture and hold their prey. If dolphins and toothed whales capture a prey individual that is too big to swallow whole, they shake it apart and swallow the pieces. This behavior has been documented several times on film, where killer whales have shaken apart a good-sized seal.

A school of small fish in tight formation. The ability of a baleen whale to open its mouth and swallow such schools is one of the factors that has led to the evolutionary success of whales.

In addition, toothed whales are thought to use sound to stun their prey. This theory was proposed by the Russian biologists Vsevolod M. Bel'kovitch and Alexei V. Yablokov in the early 1960s and was later refined by zoologist Dr. Ken Norris. Norris worked with sperm whales and recognized their ability to capture relatively small prey. He also was impressed by their ability to generate extremely loud sounds, which he believed substantiated the Russian biologists' theory about toothed whales using loud sounds to stun prey. This theory has now been applied to other toothed whales as well.

HOW MUCH DO WHALES CONSUME?

Stomachs of the large whales can hold well over 800 liters, or about 1 metric ton, of food. We can easily believe that 800 liters could represent a single meal in large cetaceans. What we do not know is how many meals a day the average whale consumes.

A right whale skim feeding at the surface. Whales skim feed by detecting a concentration of prey organisms and swimming through it with their mouths open. Sei and right whales generally use skim feeding instead of swallowing or lunge feeding.

Bubble netting is practiced by humpbacks, who blow a cloud of bubbles while swimming around a school of prey, consisting of fish or krill. The cloud of bubbles appears as a barrier to the prey, causing them to bunch up and be more easily consumed.

The humpbacks, having bunched the prey up with a bubble net, swim under the net and lunge feed on the prey. The disturbance that the bubble net makes when it reaches the surface can be seen just beyond the whales in this photo.

DO WHALES AND DOLPHINS DRINK WATER?

Whales and dolphins get their daily supply of water from their food. We humans think of being shipwrecked, drifting on a raft for days, and eventually dying of thirst because our kidneys cannot handle the salts that seawater contains. We tend to forget that humans lose a lot of water through breathing and perspiring. Even though whales breathe, they do not perspire, so they lose less water than we do. They do not drink seawater, because they do not have to. They take in a certain minimal amount of seawater with their food, but they do not actively drink it.

DO WHALES AND DOLPHINS PLAY?

Whales, like most other animals, play when they are young in order to develop the skills and reflexes that will be a very important part of their later lives. We know most about their play in captive environments. Bottlenose dolphins are frequently observed playing with objects and with other animals in captivity. They are sometimes observed mouthing feathers or coins that visitors have dropped into the tanks. Dolphins often try to keep an object away from other dolphins, teasing them with it.

Dolphins also play in the wild. Bow riding is a commonly observed behavior in which dolphins ride the bow wave of a boat. They appear to do it out of a sense of play with the boat. In reality, they are also profiting from a free ride by using the energy that the boat has created in the bow wave.

Surf riding is another activity that is popular with dolphins and even gray whales; it involves their use of natural waves (surf) in the same manner as they use waves generated by a vessel. Such activities may be correlated with travel or feeding, but usually they are not.

WHICH WHALES EXHIBIT "FRIENDLY" BEHAVIOR?

Some whales and dolphins seem to exhibit friendly behavior when they become used to nonthreatening people. There are many tales of dolphins in the wild that befriend people and appear to play with them, perhaps because they consider people to be part of their environment. However, we should remember that dolphins are still wild, and their play may easily turn into aggressive behavior.

In the early 1970s, with the increase in whale watching, a new kind of behavior occurred when gray whales approached whale-watching boats and solicited contact with people. They liked being scratched and touched. For years, few of these boats had visited the haunts of whales, and those that did had not encountered this type of behav-

A gray whale approaching whale watchers. As whale watching increased on the west coast, reports started coming in of gray whales approaching the boats. These whales were curious and allowed the whale watchers to come close and actually touch them.

ior. When the number of boats increased and the behavior began to occur, it seemed to be contagious, and since then more whales have started to behave in this way.

DO WHALES AND DOLPHINS SLEEP?

Whales do not sleep in the usual sense of profound and prolonged unconsciousness. The major reason they do not sleep soundly is that, for them, breathing is a conscious activity. When a cetacean loses consciousness, it does not breathe. Other animals, including humans, have a "breathing reflex." If we are knocked unconscious, we continue to breath automatically. Cetaceans do not do that. The fact that breathing is a conscious activity in dolphins is one of the factors that makes husbandry quite dangerous for them. In the past, if a dolphin was treated with an anesthetic prior to surgery, it stopped breathing. It was only through development of a respirator (similar to an iron lung) that surgery on dolphins became successful.

Some whales doze just under the water surface, periodically rising just enough to take a breath. In this way they can be very vulnerable to large ships, which sometimes run over whales "sleeping" in shipping lanes.

Dolphins and porpoises seem to become rejuvenated through catnaps between breaths. We now have a growing body of evidence that cetaceans sleep with one cerebral hemisphere at a time.

WHICH WHALES FORM HERDS?

Perhaps the most social whale, in terms of structured groups, is the sperm whale, which was pursued remorselessly by Yankee whalers in the middle part of the 1800s. In hunting the sperm whale, they discovered much about its social behavior. The old whale men distinguished pods, or gams, numbering up to 20 whales; schools, or shoals, numbering from 20 to 50; and herds, or bodies, of whales comprising from 50 to 100; currently, however, there is no consensus on the use of these terms.

The composition of different groups of sperm whales has a social basis. Solitary adult males are known as lone bulls. Groups of single bulls ("schoolmasters" or "proprietor bulls") with a number of females form a "harem school." Schools that consist of young males only are known as bachelor schools or forty-barrel bulls. "Nursery schools" consist of adult females and their nursing calves. Schools of pregnant females and mixed schools of juvenile males and females exist but have no group names.

Because whales are social animals, their society is organized, like those of other social animals, such as birds, deer, and humans. The basic social unit is the school,

Sperm whales socializing. These sperm whales have gotten together in a group and are obviously socializing. The group is mixed, with an adult male on his side with his mouth partly open, two calves, and two whales of intermediate size, which are probably females or immature males. It is impossible to tell exactly what is going on, but the adult male has sex on his mind, as evidenced by his erection.

pod, or gam, which consists of a relatively limited number of animals. Depending on the species and type of environment, a school may be organized for breeding, traveling (migration), group feeding, nursing and rearing the young, or any combination of these. They may change membership from time to time or retain their members and change their purpose. Sometimes schools of whales and dolphins merge into large herds. Usually the ties that originally drew these smaller schools together persist even when the larger aggregation breaks up, so that the smaller schools will reconstitute themselves.

Much is known about killer whale groups, particularly in the Pacific Northwest. These long-term groups consist of orcas that are related to one another (family groups). Each of the schools or pods has its own territory, although the territories are fluid and changing. Some specialize in feeding on salmon, and some on seals. Various groups are resident, meaning they stay in their territory for the entire year;

and some are transient, meaning they wander on a seasonal basis, falling short of migration.

One of the problems of dealing with whale groups is discerning which individuals constitute the group. Whales are primarily acoustic animals, and they may coordinate behavior over many tens of miles. As few as three or four baleen whales may cover thousands of square miles while traveling several miles apart, which makes it a bit difficult to define them as a group.

DO CETACEANS HAVE ENEMIES?

Whales are preyed upon by any animal that thinks it can succeed with an attack. Given the size of most whales, the list of predators is limited to relatively large animals. Killer whales, large sharks, and humans are the most successful predators of whales.

Whales are most subject to predation when they are young, old, or ill. When whales are very young, they usually have the protection of their mother. An orphaned or old whale is going to be subject to predation that would not normally occur if it were in the prime of life and in good health.

Questions have arisen regarding the relationship of the larger whales to swordfish. An old story that whales are preyed upon by a combination of "swordfish and threshers" was common in the eighteenth and nineteenth centuries. For many years biologists interpreted these stories as fanciful accounts of swordfish (*Xiphias*) and thresher sharks (*Alopias*) attacking whales and discounted them as mythical. Some biologists realized that one of the common names of the killer whale in German is *Schwertfisch*, referring to its dorsal fin as a sword. These accounts of "swordfish" could have been of German origin and actually referred to killer whales.

A report turned in by a ship's captain of swordfish and porpoises appearing to attack a school of large whales was investigated by the biologist S. G. Brown, who was satisfied that the captain had seen several big swordfish or billfish in a frenzy of activity around a school of whales (Brown 1960). However, he was not satisfied that there was proof that the swordfish were attacking the whales. Brown went on to hypothesize that the swordfish could have been after smaller fishes that were around the whales. We can easily see how a whale in such circumstances could accidentally be impaled by a swordfish, which could explain the finding of several swordfish swords in the bodies of blue whales.

The "thresher" is more problematic. The porpoises in the account related by Brown, which presumably were also feeding, could have given rise to the "threshers" of other accounts. Biologists Judith Perkins and Hal Whitehead have given a contemporary summary of the swordfish and thresher legends (Perkins and Whitehead 1983).

Dolphins and porpoises do not have the size advantage over predators that whales have. To protect themselves, they rely on speed and travel in schools, where, as part of a group, the individual is not as vulnerable to predation.

We also know that whales and dolphins are preyed on by the cookie-cutter shark, *Isistius brasiliensis* (see the glossary), which is a "grazing predator"—a predator that attacks large numbers of prey over a lifetime but removes only a relatively small piece of the prey, as a mosquito does. The small circular scars produced by these sharks are common on most whales and dolphins that travel through tropical waters.

HOW DO WHALES PROTECT THEMSELVES?

Much of our information about how whales protect themselves comes from many years of experience as whale predators ourselves. The whaling literature is full of accounts of the defensive action, some of it quite successful, taken by whales.

The principal weapon of most whales is their tail, or flukes. As evidenced by the many whaling boats sunk by blows of the tail, whales can actively defend themselves with that appendage. All of the axial muscles that propel a whale end in tendons that attach to the flukes. Therefore, whales are capable of bringing their flukes down with devastating force upon foes at the surface. If the foes are underwater, the blow from the flukes becomes less effective because of the resistance of the water. Underwater, the whale's principal strategy is to employ a sideways blow with its tail.

The flippers of most cetaceans do not have enough mass to be effective as defensive weapons. Only in the humpback, where the flippers commonly exceed 5 meters in length, are they of sufficient size to do damage. Despite this, to the best of our knowledge, there are no reports of humpbacks using the flippers in any aggressive behaviors.

Aside from ramming their attackers (see *Do Whales Sink Ships?*), the only remaining mode of defense is the use of their teeth. Certainly the toothed whales that have large teeth (the sperm, killer, false killer, and pygmy killer whales) use their teeth to defend themselves.

DO WHALES SINK SHIPS?

Herman Melville's novel *Moby Dick* relates the sinking of the whaling ship *Pequod*. The story of a whale ramming and sinking a vessel was not unknown in the days of whaling. It is said that Melville based at least part of his novel on the sinking of the *Essex* by a sperm whale in 1820, as told by the surviving mate, Owen Chase. Ramming is used by the smaller species of toothed whales, as evidenced by the healed bone breaks in museum specimens (see *Do Whales Fight?*).

There are interesting stories about whales sinking ships—some whose occupants had attacked whales and some who were minding their own business. One can understand that whales maddened by harpoons might attack whatever they thought was the source of the injuries. The account of the *Essex*, referred to above, is of this sort. A prominent whaling historian, Johan Tønnessen, refers to five instances where vessels were sunk by whales; in at least two, the whale concerned was a fin whale. Most of the vessels involved were catcher vessels, but one was a Danish schooner. Recent accounts of sailing vessels sunk by whales are questionable.

DO WHALES FIGHT?

Scientists sometimes define aggression as instances of conflictive behavior that are not primarily related to feeding or defense but that usually involve selection of mates or disputes over territory.

Recently bottlenose dolphins on the North Sea coast of Scotland chased and battered harbor porpoises, a much smaller species. Groups of dolphins were observed throwing porpoises up into the air. Dead harbor porpoises were found with severe premortem injuries and tooth marks with spacing that matched the bite of the dolphins. The bottlenose dolphins were clearly not attempting to eat the harbor porpoises. They may have been eliminating them as food competitors. It has also been suggested that the killing of harbor porpoises is just misplaced infanticide, which is the deliberate killing of young by their parent species. Some human societies have practiced it as a means of population control, but we do not know what the circumstances of infanticide are among dolphins.

Among many highly sexually dimorphic cetaceans, such as sperm whales and beaked whales, we have been able to see that the males have many more tooth scars than the females. These scars are of a type and spacing that indicate they have been made by other males of the same species. Because the prevalence of scars is limited to adult males, we presume they are the result of sexually competitive combat. Some beaked whales seem to have evolved with a system where the teeth erupt in males only. There is heavy ossification of the beak or rostrum, rendering it suitable as a weapon for combat with other males in the dominance ritual.

Killer whales are the only toothed whales that are known to eat other whales and dolphins. Sperm whales are larger than killer whales but have become specialized to feed on squid. Many reports exist of killer whale predation on porpoises and seals. They do frequently attack large whales, as indicated by the occurrence of killer whale tooth scars on the flukes and flippers of baleen whales. There are accounts of killer whales scavenging whale carcasses for pieces of the tongues of whales killed by whalers. In one report, killer whales cooperated with whalers in Eden in New South Wales, Australia, to signal the presence of baleen whales in the

A group of at least three male narwhals with their tusks out of the water. The tip of the closest narwhal's tusk is broken off, and its pulp cavity may have been filled by the tip of another narwhal's tusk.

vicinity. They then scavenged the carcasses after the whalers killed their prey. Studies of stranded gray, blue, fin, and minke whales in Alaska have shown killer whale predation to be in evidence.

HOW AND WHY DO NARWHALS USE THEIR TUSKS?

Numerous theories have been presented about the narwhal's use of its tusk. All narwhals are equipped with a pair of teeth, one on each side of the upper jaw. All lack teeth in the lower jaw. Normally, only the left tooth erupts in adult males and becomes a tusk. The structure of the tusk is unusual in that it grows in a sinistral spiral, giving it the shape of an enormous left-hand screw. Occasionally (on the order of 1 percent of the population), both tusks erupt in a narwhal. A number of two-tusked individuals are recorded as females. A curious thing about the two-tusked narwhals is that the tusks are not symmetrically spiraled. You would think if the left tusk has a left-hand spiral then the right tusk would have a right-hand spiral, but both tusks are sinistral.

An early theory of tusk usage was that the narwhal used its tusk as a spear to capture prey or perhaps to dislodge bottom fish from their crevices in the rocks. No observations have been recorded to verify this theory.

Morton Porsild, an observer who studied museum specimens of narwhal tusks in Europe, found that an appreciable number were broken during the life of the narwhal (107 broken out of a total of 314 observed, or 34 percent). He also found that four tips of the tusks had been plugged. He extracted the plugs and found them to be the tip of another narwhal's tusk. The pulp cavity extends nearly to the tip, and if the break were to expose the pulp, it would be extremely painful. Porsild believed that at least four narwhals with such breaks managed to get another narwhal to insert the tip of its tusk into the pulp cavity and break it away (Porsild 1922). This is a delicate operation, like threading a needle, so both animals would have to be aware of what is going on. If this is indeed how the tusks were plugged, it is one of the most altruistic behaviors of whales.

Like many other animals with such sexually dimorphic characters, the narwhal may use its tusk in the same way as a deer uses its antlers: to determine mating success. This theory is substantiated with an increasing number of behavioral observations of narwhals seen crossing their tusks above the water line.

DO WHALES AND DOLPHINS HAVE VOICES?

Whalers and fishermen have known for centuries that whales produce sound, and it is logical to assume that any animal that produces sound is capable of perceiving sound. Biologists William Schevill and Barbara Lawrence demonstrated that the bottlenose dolphin can hear sound frequencies up to 100 kilohertz (Schevill and Lawrence 1949). It was not until Winthrop Kellogg published his research in 1961 that we really began to understand dolphin echolocation. Dolphins can hear the range of frequencies that we hear, as well as higher frequencies up to and in excess of 150 kilohertz. That is about the upper range of hearing among bats.

Our best data are for animals that have been kept in captivity, which eliminates most of the larger whales. Baleen whales produce sounds at a much lower frequency than that of dolphins. In fact, blue whales produce powerful sounds that are below the human hearing range. Scientists in 1971 recorded sounds from a blue whale that extended down to 12.5 hertz and were "the most powerful sustained utterances known from whales or any other living source" (Cummings and Thompson 1971:

1193). Biologists have computed that fin whales using 10-hertz sounds can communicate over 800 kilometers. Sounds that range in frequency from 40 hertz to 5 kilohertz have been recorded from humpbacks.

HOW DO WHALES AND DOLPHINS MAKE SOUNDS?

The question of how whales produce sounds and where those sounds are produced in the body is complicated. Whales do not have vocal cords in the same anatomical sense as humans do. In fact, the structures known as vocal cords in humans are rarely found in other animals. However, dogs, cats, horses, and most other non-human animals lacking vocal cords do have "voices"; they have structures known as vocal folds.

Animals produce sound in one of two primary ways: either they cause air to pass over vocal cords or folds so that they vibrate, or they cause air to fill a cavity so that it resonates. We are familiar with the first of these two mechanisms as it relates to our own voices. Hold your fingers against your larynx while speaking and feel the vibration of the vocal cords. Whistles illustrate the second mechanism, where air is blown into a cavity, causing it to resonate.

Whales and dolphins have a complicated respiratory system that is full of structures that can provide for both of these mechanisms. It has been hypothesized that dolphin whistles are generated in their larynges and that the high-frequency sounds that dolphins use for echolocation are made by complex structures in their foreheads. We hear this latter type of sound not as high-frequency sounds but as clicks. Trained dolphins produce many air-borne sounds with their mouths and blowholes much in the manner of "Bronx cheers" or "raspberries." Sperm whales also have very complex sounds that scientists think are produced in their foreheads, which are the size of a small freight car.

The sounds that dolphins and porpoises make can be divided into two categories, "whistles" and "clicks." Whistles are relatively low-frequency sounds, within the range of human hearing, and are mainly used for communication between dolphins. Clicks are broad-band sounds, extending into high frequencies beyond the range that humans can hear directly. They are emitted in bursts called pulse trains. If a dolphin emits pulse trains consisting of bursts of sound that have a frequency of 100,000 hertz, turning on and off 2,000 times a second, those vocalizations are heard by humans as a ragged or variable 2,000-hertz tone (the highest C on a piano keyboard is 2,112 hertz). The largest odontocete, the sperm whale, emits only clicks.

Baleen whales make many sounds that are within the human limits of hearing. Those sounds are probably produced by the larynx, a very large complicated apparatus. The songs of humpback whales fall into this category. Baleen whales also produce sounds that are far below the human hearing range. These sounds are probably produced by a combination of the larynx and the resonant qualities of the bony nasal passage.

Baleen whales do not emit high-frequency sounds that are in the same category as the dolphin clicks. All of the sounds that baleen whales produce are in the human hearing range or below, the part of the range where the sounds carry farther in the water.

HOW DO CETACEANS USE SONAR?

The ability of animals to vocalize with the purpose of listening to the echoes of their sounds is known as echolocation, or sonar (*sound navigation and ranging*). Sonar can be active, when cetaceans listen to sounds they have made (echolocation), or passive, when they listen to echoes of sounds from other sources. The toothed whales make sounds high enough in frequency to produce details that are comparable to the detailed data that we perceive through our vision. They can "see" prey and obstacles in the water. In addition they can "see" into other animals in the same way that doctors use ultrasound to look into the human body. It is possible that large whales use the echoes from the low-frequency sounds they produce to detect ice openings or to visualize the shoreline.

WHY DO SOME WHALES SING?

Whale songs are a series of different themes given in a predictable order. The most famous songs are the vocalizations produced by the humpback whale. Some other cetacean sounds, such as the sounds produced by the beluga, are not really songs in

A cetologist onboard the research vessel listens to the vocalizations produced by the humpback whale in the photo on page 88. (Photo obtained under NMFS permit #987)

A humpback "singing" into a hydrophone (an underwater microphone), seen just ahead of the whale. The flippers and jaw of the whale are covered with circular scars left by barnacles. (Photo obtained under NMFS permit #987)

the strict sense. Fishermen heard them through the boat hull and likened their tonality to the singing of canaries, thus the nickname "sea canaries." Whale songs, like bird songs, serve as a means of communication. Only male humpbacks actually sing; it is one of the most noticeable means of advertising for a mate.

DO WHALES AND DOLPHINS COMMUNICATE WITH ONE ANOTHER?

Most animals communicate, even if it is just simple expression of physical wants, such as hunger. Whales and dolphins, as social mammals, communicate constantly. Humans think of communication as a process that is primarily verbal, but cetaceans are highly vocal animals in a nonverbal sense. Some of cetaceans' vocalization serves in echolocation (or sonar), but most of it serves for communication. An animal need not have a language to communicate verbally; for instance, consider how dogs are able to communicate that they are hungry or want to play or go outside. By just making a noise while it is eating, an animal can communicate to others that (1) it is in a particular place, (2) it has successfully captured prey, (3) it is consuming that prey, and (4) it does not want anyone else to eat its meal, or it wants its friends to come and share its meal.

Some work with dolphin vocalizations has determined that they have "signature whistles," which means that one dolphin can recognize another individual dolphin by its whistle. Other work done with dolphins involved separating them into different pools that were acoustically linked. Dolphins in one pool were given instructions for a task. They had to communicate that information acoustically to dolphins they couldn't see in the other pool. Then the dolphins in the second pool had to press a paddle and communicate the results to the trainers, which they were able to do.

DO CETACEANS COMMUNICATE BY ANYTHING EXCEPT SOUND?

In addition to communicating acoustically, cetaceans also communicate visually. Vision is somewhat less useful underwater, but you have only to look at dolphins following the visual cue of a trainer in order to realize that these animals are highly visually oriented.

Cetaceans also communicate through touch. They have tactilely sensitive skin and are thigmotactic (they enjoy touching one another). A prime example of this is the tactile communication that passes between a mother and her calf.

We do not yet understand the chemical senses of cetaceans (taste and smell). In future years we will probably discover dolphin "scents" and realize that they communicate chemically as well.

Much of the early information on dolphin vocalization and group behavior was acquired by working with captive animals. Admittedly, behavioral studies on captive animals have some limitations, but for many studies this is the only feasible way to investigate dolphins closely, because the situation can be controlled much more effectively than in the wild. Captive studies have helped us accomplish much work on communication in dolphins and the concept of a "dolphin language."

CAN THE LARGER WHALES BE TRAINED?

Our inability to maintain large whales in captivity limits our opportunities to train them. However, gray whale calves have been kept in captivity in California; pilot

A group of killer whales that have been trained to jump in unison at an oceanarium. Even among the larger species, such training is common in cetaceans, particularly behaviors that are impressive to the public.

whales and killer whales (which are technically both large dolphins) are kept in captivity in many oceanaria; and minke whales have been confined in bays in Japan. These animals have responded well to training. Whales have been trained to leap on command, carry trainers around on their backs, squirt a mouthful of water at people, and wave their flippers in the air. Of course, the larger an animal is, the more expensive it is to train.

WHAT CAN BE DONE WITH A SICK WHALE?

The decision to rehabilitate sick cetaceans—that is, to keep them for a period of time while they are medicated and then release them into the wild—largely depends on available money, maintenance facilities, and the nature of the sickness. Usually the animals chosen are young, because it is difficult to transport and care for adults. They must be handled, diagnosed, medicated, and fed, which is tricky and expensive. In essence, they become acclimated to one extent or another to captivity. If the disease symptoms finally abate but the animals still show blood antibodies that indicate disease organisms may yet be present, should they be released into the wild, possibly serving as a disease carrier? Could they have been exposed to other diseases while in captivity? If we keep them a long time, their antibodies will subside, but we run the risk of the animals' becoming accustomed to captivity and perhaps losing their natural fear of humans.

Cases where cetaceans have been successfully released from captivity are rare. Usually they are rehabilitated until their acute disease symptoms have abated and then are returned to the sea without any follow-up monitoring. We just do not know what becomes of these animals. Some cetaceans that have been released with radio and satellite tags, enabling researchers to locate and monitor them, have been incorporated into schools and have successfully returned to a normal life.

WHY DO WHALES STRAND?

Single strandings of cetaceans are considered the result of normal mortality, unassociated with human actions; there is nothing mysterious about them. Regardless of whether the cetaceans are alive or dead when they strand, they soon die.

Mass strandings occur when a group or school of cetaceans comes ashore alive. They rapidly encounter serious problems with desiccation, sunburn, and other aspects of exposure. We know that group cohesiveness is certainly a factor in mass strandings, because species that mass strand have very tight schools, although so do many species that do not mass strand. Basically, our problem in understanding mass stranding is a lack of knowledge about the mechanisms that result in movements of whales into areas where they strand. Some possible motivations are behavioral, such as the school following an impaired leader. Cetaceans can have a variety of diseases, some acute and some chronic. Some species are more prone to mass strand than others, such as pilot whales. There is a theory that anomalies in the earth's magnetic field may interfere with whales' navigation and cause mass strandings. After years of studying stranded animals, we still do not know why they do it.

Dead stranded animals represent not only a source of data on what factors were responsible for the stranding but also invaluable data on the natural history and other characteristics of the species involved. Scientists can obtain data on the age, reproductive condition, food habits, parasites, chronic diseases, and pollutant levels from individual strandings. Strandings also provide us with a source of specimens for anatomical and systematic investigations.

Ventral view of a dead southern right whale, stranded along the coast of Argentina. The scale provided by the researcher balancing on the fin indicates that this whale is a large adult.

WHAT SHOULD BE DONE WHEN A CETACEAN STRANDING OCCURS?

When a cetacean strands, the local stranding network should be notified. The network is in contact with the U.S. Coast Guard, the National Park Service, and state and local law enforcement services. Contacting any one of those agencies should get information to the stranding network.

Until persons familiar with stranding arrive on the scene, you can try make the animal comfortable if it is alive. Cetaceans are extremely sensitive to the sun and rapidly suffer from sunburn, even on a cloudy day in the winter. Cover the animal with something to shield it from the sun—towels or sheets or anything that will not abrade the skin. Cetaceans are also susceptible to overheating, which is especially likely when they are dry. Wetting the animal is a good way to ensure that its normal thermoregulatory mechanisms can work. Pay particular attention to keeping the flippers and tail wet, because they radiate most of the heat and need to be kept cool to do that efficiently.

Cetaceans are uncomfortable on anything but a sandy beach. If the flippers get pinned under the animal and circulation ceases in them, they will become gangrenous. If you can, excavate cavities in the sand, so that the flippers hang freely.

If the animal is small and near a sheltered body of water, you might think about rolling it onto a tarp and pulling the animal by the tarp into the water. Try not to pull on the animal directly, because you could aggravate the injuries that it might already have experienced. Use caution, because cetaceans can be very heavy. Also, cetaceans are very powerful, and even sick dolphins can inadvertently injure the persons trying to care for them. Be particularly careful of the flukes.

Depending on the medical condition of the stranded animals and the facilities at the disposal of the stranding personnel, it may be possible to take some individuals to rehabilitation facilities and attempt to save them.

.2.

WHALE EVOLUTION AND DIVERSITY

WHAT ARE THE EARLIEST WHALE FOSSILS?

The cetacean fossil record is extremely fragmentary, so our current scientific theories, which are based on it, will probably change as new finds are made. Current evidence suggests that whales evolved from animals resembling large hyenas within the now-extinct terrestrial mammalian family Mesonychidae. The mesonychids were members of the order Creodonta and were probably generalized carnivores/omnivores eating meat, fish, vegetation, and whatever else the land and shallow waters provided. The creodonts were of the primitive mammal groups that gave rise to artiodactyls, even-toed hoofed mammals such as sheep and cows.

The first whales are called archaeocetes. The term *archaeocete* derives from the Greek words *archaios* (ancient) and *ketos* (whale). The oldest known whale was found in sediments in Pakistan; it is known as *Pakicetus* and is thought to have lived approximately 50 million years ago. A number of related forms have also been found but in slightly younger sediments—*Ambulocetus*, *Rodhocetus*, and *Indocetus* in Pakistan and *Protocetus* in Egypt. These whales appear to have had functional hind limbs and were probably amphibious. By the early Oligocene, 40 million years ago, the baleen whales (mysticetes) split off from the forms that evolved into the modern toothed whales (odontocetes). The first baleen whales had teeth and did not develop baleen until later.

The early fossil whales differed from living whales in a variety of ways. The most distinguishing feature of archaeocetes was their fully differentiated (heterodont) dentition; that is, they had incisors, canines, premolars, and molars. Living odontocete whales have simplified teeth that resemble incisors (homodont), and living baleen whales have no teeth at all. Archaeocetes also differed in many features of the skull.

Killer whale, *Orcinus orca*.

95

Archaeocetes are sometimes typified by the genus *Basilosaurus* (also known as zeuglodon). The members of this genus were extremely long-bodied, reaching upward of 16 meters, appearing slender and snakelike. Most scientists agree that *Basilosaurus* was not ancestral to living whales and was unusual in body form. A different but related archaeocete, *Zygorhiza,* with a body form like a large dolphin's, probably gave rise to modern whales.

For many years scientists thought that diversity among the archaeocetes was slight. Recent finds indicate that a major evolutionary divergence of structure and habitation (a radiation) occurred when whales first took to the water. We are just beginning to appreciate the magnitude of the archaeocete radiation. They evolved into many different forms and lived in many different habitats.

WHEN DID THE FIRST WHALES LIVE?

Dating of fossil whales involves either dating the fossils directly or dating the beds in which the fossils are found. For relatively recent fossils, up about to 20,000 years old, a specimen can be dated by radiocarbon techniques. The accuracy varies and is worst for very young and very old samples. For instance, it is not unusual to submit a sample for dating and receive results indicating it is 300 years old, plus or minus 500 years, which means it could be 800 years old or not yet born (coming from 200 years in the future).

In most cases we try to calculate the age of the sediments in which the fossil occurs. A common way of dating the sediments is to correlate them with igneous rocks, such as those in a lava flow. Igneous rocks can be absolutely dated by the ratio of the unstable isotope potassium 40 to the stable element into which it changes, argon 40. These analyses are subject to variation, and the difficulties in performing them are compounded by the physical arrangement of the sedimentary and the igneous beds. Because fossils are not found in igneous rock, two layers of lava are needed, one below and one above the fossil layers. Nature does not always provide us with such handy dating tools.

The point to remember is that, with current techniques, determining the age of a fossil may be imprecise, but even approximations are useful. Often, finding that a fossil is older than one date or younger than another in the best that can be done.

WERE THE FOSSIL WHALES REALLY DIFFERENT FROM MODERN WHALES?

The earliest whales certainly differed from the modern whales in that they were more generalized biologically. They had not had time to develop the specializations that typify their modern descendants. As with many evolutionary breakthroughs, the reentry of cetaceans into ocean waters produced an explosion of different ways to exploit the marine environment. For a time, baleen whales had numerous teeth as well as baleen. The diversity of whales peaked in the Miocene, roughly 20 million years ago.

The fossil record has given us tantalizing glimpses of the total diversity of fossil whales. One such glimpse involves a fossil from the early Pliocene deposits, about 4 million years old, along the coast of Peru. This unique fossil is known as *Odobeno-cetops* (a whale that resembled a walrus), and it was sufficiently strange that its acceptance as a cetacean was doubted in several quarters. It was about 3 to 4 meters in length and carried one tusk that was directed toward the rear, like those of the walrus. The skull was radically different from other whales, but there were enough similarities to modern narwhals and dolphins that it is classified in the same group as they are (superfamily Delphinoidea).

WHAT CHANGES THE NUMBER OF WHALE SPECIES?

Additions to the list of cetacean species take place when new species are discovered and named on the basis of their description and when we rename previously discovered species that were once thought to be synonyms but are now recognized as different from one another. Deletions to the list take place when we drop a name because we realize that two species are actually not different but the same, or synonymous. At the time of this book's publication, 656 specific names are applied to cetaceans. We now think this list of names is larger than it should be, because we know that it was based on incomplete species descriptions that do not account for the variation that we now know occurs within a single species. Taxonomists use the date of Linnaeus's 10th edition of *Systema Naturae* (1758) as a starting point to limit the number of names. They do not consider names that were employed before 1758 as valid. It was another 100 years later, with the publication of Darwin's theory of the origin of species by natural selection, before we first began to understand variation within species. And not until the turn of the twentieth century did we begin to understand the extent of this variation.

Some of the species that have been recently recognized as valid include Fraser's dolphin, *Lagenodelphis hosei,* and the Clymene dolphin, *Stenella clymene*. Now that statistically usable samples are being compiled for many species, it should be possible for us to recognize previously named species that the scientific community has not recognized (cryptic species) and certain subspecies. Refinement of cetacean diversity is taking place slowly. The concept of subspecies and population differences is just beginning to be utilized.

Blue whale.

Fin whale.

Humpback whale.

Right whale.

Bowhead whale.

Gray whale.

Sperm whale.

Narwhal.

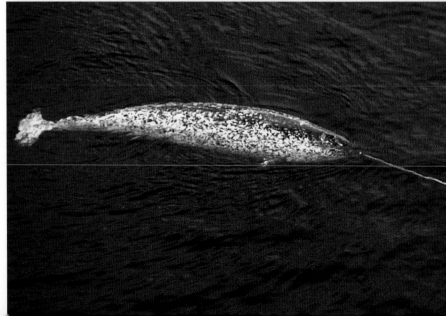

Belugas.

Bottlenose whale.

False killer whales.

Pilot whales.

Killer whale.

Hector's dolphins.

Spotted dolphin.

Dusky dolphin.

Risso's dolphin.

Rough-toothed dolphin.

Common dolphins.

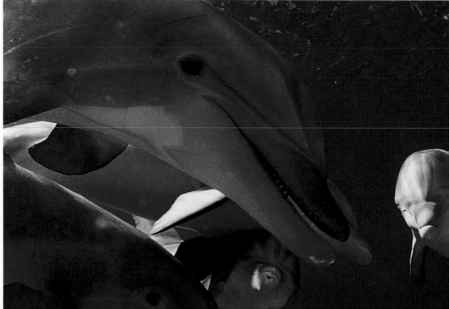

Bottlenose dolphins.

Spinner dolphin.

Amazon river dolphin.

A recent species to be described was the pygmy beaked whale, *Mesoplodon peruvianus*, in 1991, from Peru and subsequently recognized also off the coast of Mexico. The last new species to be described was *Mesoplodon bahamondi* (Bahamonde's beaked whale), a species named in 1995 on the basis of a partial skull from the Juan Fernández Islands, off Chile.

Robert Pitman and others wrote about the existence of an unnamed beaked whale in the tropical Pacific in 1987. It is just a matter of time until specimens of that species are collected and the whale is named. The bulk of the undescribed species are in little-known open-ocean (pelagic) groups, such as the beaked whales, and are in parts of the world where collecting is infrequent, such as the tropics.

If we compare the present cetacean diversity with the fossil record, incomplete though it may be, we find representatives of every modern cetacean group in the Miocene (20 million years ago) as well as many groups that have since become extinct. Cetacean diversity appears to be naturally diminishing.

The living whales, which were thought to number 300 species in the late 1800s, are now thought to consist of fewer than 100 species. We recognize 86 in our list in Appendix 1. The following are species accounts consisting of examples of the major types of whales.

BALEEN WHALES (SUBORDER MYSTICETI)

Right Whales (Family Balaenidae)

Southern Right Whale (*Eubalaena australis*), North Atlantic Right Whale (*Eubalaena glacialis*), North Pacific Right Whale (*Eubalaena japonica*), and Bowhead, or Greenland Right Whale (*Balaena mysticetus*)   The right whale was the ubiquitous whale in much of classic literature. It formed the basis of early whaling, and, together with the bowhead, it supplied the world with baleen. It is said that the name "right whale" is derived from its being the "right," or correct, whale to kill. Its coastal habits put it in the range of many early whalers; it had abundant and valuable baleen and blubber; and it floated when dead, making it easier to work with. Its one drawback was a tendency to turn on the harpooning boat and sink it.

During the nineteenth century, scientists debated how many species of right whale there were. At that time, they had not seen any specimens of the Greenland right whale, or bowhead (*Balaena mysticetus*), and some of them thought that it was just a more northerly population of the common black, or Biscayan, right whale. It was not until Daniel F. Eschricht and Johannes T. Reinhardt published a detailed description of the Greenland right whale in 1861 that scientists realized that more than one species existed. The debate still goes on, with scientists disagreeing about

The nose of a right whale, seen here in a view from the front, is a unique complex of skin features that are known as callosities. The white in the view is made up of hundreds of whale lice that cluster on the dark gray, wartlike surface of the callosity.

whether one species of black right whale (*Eubalaena glacialis*) exists with a worldwide distribution or whether three species exist, one in the North Atlantic (*Eubalaena glacialis*), one in the North Pacific (*E. japonica*), and one in the Southern Hemisphere (*E. australis*). There seems to be a consensus that the three populations are separate; the disagreement is over whether they have been separated long enough to be recognized as separate species.

The North Atlantic right whale is about 15 meters long. It is characterized by a large head, as much as a third of the body length, with an arching upper jaw that supports up to 390 baleen plates on each side. The baleen plates average 2.1 meters in length in an adult. Only the bowhead has larger plates, up to 4.2 meters long. The North Atlantic right whale is remarkably rotund. Its maximum diameter (between the flippers and the navel) is slightly better than half of its total length. All right whales lack a dorsal fin and have paddle-shaped flippers that are about 2.4 meters long in adults. The flukes are very broad, about 5.4 meters wide.

Early in its existence, the North Atlantic right whale developed rough patches of skin, similar to warts, on its head and jaws. These discrete patches are known as callosities and are home to enormous numbers of whale lice (cyamids). Because these callosities appear white, they were confused with barnacle patches, which are common on humpbacks. In fact, the skin of the callosities is dark gray; the white color comes from the cyamids. Only one rare genus of barnacle (*Tubicinella*) has been documented as occurring on right whales, and it burrows into the skin of the back.

Right whales live in the temperate waters of the world's oceans. They are absent in the tropics and in the Arctic, although there have been sightings along the tip of the Antarctic Peninsula. Migration occurs between summer feeding grounds and winter breeding grounds. Although they have been protected against whaling, populations have not come back. The North Pacific right whales are the rarest. Because of their callosities, individual right whales were recognizable early on, and catalogs have been assembled for right whales off the northeast coast of North America and the coasts of South Africa, Patagonia, and southern Australia.

On the basis of the North Atlantic population, we know that female right whales reach sexual maturity at 9 to 10 years, and mature females reproduce on the average of once every 3.67 years. The calves at birth are around 4.4 meters long. Mature right whales reach a length of about 17 meters.

Right whales have the distinction of having the world record for the largest testes. A right whale taken by the Japanese off Kodiak Island, Alaska, in August 1961 had a combined testes weight of over 972 kilograms. The largest testis was 201 centimeters long and 78 centimeters in diameter.

Right whales are skim feeders, preying almost exclusively on minute crustaceans known as copepods.

Pygmy Right Whale (Family Neobalaenidae)

Pygmy Right Whale (*Caperea marginata*)   One of the least-known whales, certainly the least known of the baleen whales, is the pygmy right whale. It is known primarily from stranded animals, although the Russians took a few in the Antarctic for scientific purposes.

Caperea is from the Latin *caperro*, "to wrinkle," an allusion to the wrinkled appearance of the tympanic bulla. *Marginata* is from the Latin *margino*, meaning "to provide with a border," an allusion to the dark border of the baleen.

The pygmy right whale, as its common name implies, is the smallest of the baleen whales. The longest ones known were a 6.1-meter male that stranded in Cloudy Bay, Australia, and a 6.45-meter female that stranded in Stanley, Australia.

The species is dark gray or black on its back, shading into white on the belly. The snout is dark gray, contrasting with the gums, which are white. The lower jaw is dark to light gray, merging to a paler throat area. The baleen plates are white to yellow with a dark outer margin and white to yellow bristles. The flippers are relatively small and paddle shaped. They are dark above and light below, sometimes with a darker tip and leading edge. The dorsal fin is small, sickle-shaped, and located about two-thirds to three-quarters of the distance from the snout to the flukes. The flukes are black above and light below, with a dark band along the ventral surface of the posterior edge.

One of the most striking things about the pygmy right whale is its enormous chest. The body in all mammals is divided into cervical (neck), thoracic (chest), lumbar (abdomen), sacral (hip, pelvic), and caudal regions, according to type of vertebra. Nearly all mammals have 7 cervical vertebrae, and most whales and dolphins have 12 to 15 thoracic vertebrae with their associated ribs and about the same number, 12 to 15, of lumbar vertebrae. Cetaceans lack the connection of the pelvis to the vertebral column; so they do not have sacral vertebrae. Their caudal vertebrae number about 20 to 30. Pygmy right whales also have 7 cervical vertebrae, but they have 18 thoracic and only 2 lumbar and 16 caudal vertebrae. These extremely long thoracic and short lumbar regions result in a stiffer body and may account for the tales of the pygmy right whale's strange method of swimming, in which it moves its head up and down in synchrony with its fluke.

Pygmy right whales possess an unusually large laryngeal sac ventral to the larynx. The sac was found to be 50 centimeters long in a 5.9-meter adult male. We think the sac plays a part in vocalizations. A series of short thumplike pulses of low-frequency sound (60 to 120 hertz) was recorded from a juvenile pygmy right whale in Portland Harbor, Victoria, Australia.

The pygmy right whale is restricted to the cold-temperate waters of the Southern Hemisphere north of the Antarctic Convergence. Because of the difficulty of recognizing it at sea, most of the records are from strandings, which have been discovered mainly from the coasts of southern Australia and New Zealand. Scattered strandings have occurred in South Africa, Argentina, the Falklands, and the Crozet Islands, in the Indian Ocean. A few were captured by Russian whalers in the South Atlantic.

This species was first known to science for many years as *Balaena marginata* on the basis of an 1846 description of three plates of baleen, 50 centimeters long and 6.4 centimeters wide at the base, from Western Australia. This is rather narrow baleen, similar in shape to the baleen of right whales; so the pygmy right whale, which had been commonly named the Western Australian whale, was originally given the generic name *Balaena* (the original generic name of the right whale).

In the meantime further discoveries in New Zealand resulted in the 1864 naming of *Caperea antipodarum* on the basis of a right earbone (bulla) from the Otago Peninsula and verbal descriptions of southern right whales.

In 1870 a skull, with baleen associated, came to the attention of a British scientist, John Edward Gray, and demonstrated that the baleen of "*Balaena marginata*" was from an animal that was considered, on the basis of its skull, not to be the genus *Balaena*. Gray then formed a new generic name, *Neobalaena* (*neo* = new + *balaena* = whale), to accommodate that whale, which then became *Neobalaena marginata*.

By 1873 this whale's vernacular name had been changed to "pygmy right whale" on the basis of its small size (5 meters) and the supposed relationship of its baleen to that of the other right whales.

John Gray died in 1875. In 1885, when a new list of the specimens of cetaceans in the British Museum was being compiled, scientists realized that the specimen bearing the name *Caperea* was identical to the specimen named *Neobalaena*. The latter name (*Neobalaena*) was now chosen because it was based on one species, whereas *Caperea* was based on a combination of two. *Neobalaena* was then used for many years as the generic name of the pygmy right whale. Historians later realized that *Caperea* was the older of the two names and thus had priority under the International Code of Zoological Nomenclature.

For many years, the pygmy right whale was still considered by most cetacean taxonomists to be in the same family as the other right whales (Balaenidae). In 1984 two paleontologists who were considering the relationships of the gray whale to other cetaceans did a detailed comparison of the pygmy right whale to the right whale and determined that the pygmy was sufficiently different to warrant a family of its own. Fortunately, there was a family name around, the Neobalaenidae, which had been proposed in 1873 by Gray to house *Neobalaena marginata*.

Gray Whale (Family Eschrichtiidae)

Gray Whale (*Eschrichtius robustus*)   The gray whale is the sole member of the family Eschrichtiidae. The name *Eschrichtius* is in honor of a nineteenth-century Danish zoologist, Daniel Eschricht; *robustus* means "oaken" or "strong." And strong

A gray whale with mud in its mouth. Gray whales are one of the few cetaceans that normally feed on the bottom. Here a gray whale has just picked up a mouthful of silt to strain it through its baleen, retaining the animals that it is interested in feeding on in its mouth.

it is, traveling 18,000 kilometers roundtrip on the longest migration of any whale or, indeed, of any mammal. For nineteenth-century whalers, this whale was so strong and rambunctious that each attempt to capture a gray became a dangerous confrontation. Commonly called devil fish, hard head, and mud digger, it slammed and rammed boats, changed course, zigzagged as it tried to escape, and killed and wounded many of the whalers before it was killed. Nevertheless, in some three decades, the gray whale was almost completely annihilated by whalers anxious to fill their barrels with oil and their pockets with cash. By the turn of the century, most naturalists thought the gray whale was extinct. The story of the gray whale in the eastern North Pacific is one of a tenacious hold on life, a story of a rebounding population after near extinction, and one of the few successes in the conservation of whales.

Easily recognized, gray whales are a mottled gray with splotches of white. Their entire bodies, particularly their heads, are encrusted with clusters of barnacles and orange or yellow whale lice (cyamids). Even calves, within a few days of their birth, acquire these parasites that live in indentations or creases on the whale's body and never leave unless there is some contact with another gray whale, as in mating. The gray whale's head is narrow and from above appears to be triangular; the mouth is slightly arched. On the jaws are many small sensitive hairs; the throat carries only two to four ventral creases or grooves, not nearly as long or abundant as the throat grooves or pleats in many other baleen whales. The baleen is cream to light yellow in color, coarse, and thicker than that of other baleen whales, and the baleen plates are fewer in number (140 to 180 on each side of the upper jaw) compared with those of other baleen whales. Gray whales lack a dorsal fin, present on most other whales. Instead, they have a series of low knoblike ridges on their back, starting where the dorsal fin is located on other whales. There are only four digits (fingers) in the flippers; the first finger (thumb) is absent. Their distinctive appearance and habit of swimming close to shore has been a bonanza for the whale-watching industry, which looks forward to their migration each year when the whales ply the waters between the Arctic seas, where they feed in the summer, and the shallow lagoons of the Baja California peninsula, where they calve and mate in the winter.

Two stocks of gray whales exist today: the eastern North Pacific, or California, stock and the western North Pacific, or Korean, stock. The latter is close to extinction, having been reduced to about 100 individuals. Both fossil remains and published records indicate that a third stock of gray whales lived in the North Atlantic within historic times. As recently as the early eighteenth century, whalers in New England hunted whales they called scrag whales. The description of such animals agrees with gray whale diagnostic characteristics, as do subfossil skeletal parts found in the western Atlantic from New Jersey to Florida and the eastern Atlantic from the coasts of Sweden, the Netherlands, Belgium, and England. Gray whales in the North Atlantic became extinct; those who knew the whale liked to eat it. The species name, *Eschrichtius robustus,* is based on subfossil Atlantic specimens.

The California stock spends the summer, May through September, mainly in the northern and western Bering Sea, the Chukchi Sea, and the western Beaufort Sea, feeding in the shallow waters of the continental shelf. A small population also exists on the west coast of Vancouver Island, and a few individuals occur along the coast southward to central California. Gray whales are not found in the deep waters of the southwestern Bering Sea, because they are primarily bottom-feeders, mainly eating various species of amphipods (small crustaceans) and, only secondarily, other invertebrates. Russian researchers analyzed stomach samples from whales taken off the coast of the Chukchi Peninsula between 1965 and 1969 and in 1979 and 1980 and found prey items including 60 species of amphipods and 80 to 90 other invertebrates, in addition to algae, sand, silt, and gravel (Yablokov and Bogoslovskaya 1984). Frequently, only one or two species of amphipods make up 90 percent of the food remains; other species occur in small numbers. Russian scientists estimate that members of the California population in these northern waters consume 850,000 metric tons of food species in one season.

Some researchers believe the gray stirs up the bottom with its snout, taking in prey items plus sediment, then filters out the water and expels the sediment through the baleen and in that way causes mud plumes or a sediment trail; others think feeding is more a sucking action, using the tongue and lips to separate the sediment. It is possibly both. The gray is also able to feed by skimming and engulfing pelagic prey, and occasionally it will feed on spawning squid or small bait fishes. The gray whale favors its right side while feeding. Researchers have observed greater wear on the baleen on the right side of the head; also, fewer barnacles and more skin abrasions occur on that side.

In November, the migration south begins and continues through December. The Arctic population migrates through the Unimak Pass, in the Aleutians, hugs the coast around the Gulf of Alaska (they are even seen in the surf zone), and strikes out for the shallow protected lagoons of Baja California, passing very close to the coast at Monterey and Point Sur, California. Most of the calves are born in the areas of Laguna Guerro Negro, Laguna Ojo de Liebre (known as Scammon's Lagoon), Laguna San Ignacio, Bahía San Juanico, Estero de Soledad, and Bahía Magdalena, to name but a few. The first whales to migrate south are the pregnant females, followed by females recently ovulating, then immature females and adult males, and finally immature males. Calves are born from January through March. Gray whales give birth every two years.

The northern migration route again is coastal. Newly pregnant females travel first, then females without calves, then adult males, then immatures. This time the females with nursing calves return last, sometimes in May. By the age of seven months the calves are weaned and ready to eat by themselves.

Female gray whales at physical maturity are generally larger than males of the same age. At age 40, females are about 14.1 meters long; males, 13.0 meters. At the average age of 8 years (or between 5 and 11 years), the whales become sexually mature, measuring an average of 11.1 meters for males and 11.7 meters for females. Weights are quite impressive: a male 11.72 meters long weighed 15,686 kilograms; another, at 12.4 meters, weighed 16,594 kilograms; a pregnant female, at 13.35 meters in length, weighed 31,466 kilograms! Newborn calves about 4.6 meters long weigh about 500 kilograms.

Other than man, killer whales are the only animals that attack and kill gray whales. They bite their pectoral flippers, flukes, and even try to open the whale's mouth to bite the tongue. Young calves are at greatest risk on migration, but many grays that escape carry tooth scars inflicted by killer whales.

Gray whale reproductive behavior includes courting activities, where both males and females arch out of the water, roll around the body axis, and swim in line, and mating behavior, where the whale may swim in circles, swim on its side with one flipper held above the water, and engage in touching displays and other exaggerated positions. Females, sometimes with calves alongside, will be accompanied by a male intent on mating. A second bull often follows the two in readiness to mate if he has the chance. This is a dangerous situation for the female. There is much rolling about by the two competitors, and often the female will try to escape and protect her calf. She can lie belly up so that the males cannot reach her genitals, but finally one of the males rotates his body under hers so that when she turns over to breathe, copulation can take place.

"Friendly" or curious behavior was previously a rarity for a whale noted for its aggressive defense. In 1975 it was first noticed that gray whales approached and allowed passengers in whale-watching boats and skiffs to pet them. From that time to the present, this phenomenon has increased. Whales of both sexes and all ages seem to be attracted to small skiffs with their engines idling. When the engines are shut off, the whales frequently leave. Sometimes they remain close to a boat or follow it when it is moving slowly. They may push the skiff or jostle it, sometimes hard enough to knock off a passenger or two. Whale watchers have reported that gray whales produce a sound in the air similar to a Bronx cheer. It is possible that these whales now frequently approach whale-watching vessels because they have become accustomed to many of these vessels in some of the lagoons or because they are no longer harassed by whalers or sport fishermen. Whatever the reason, the uncommon has become a common event, with most tourist boats experiencing encounters of a friendly nature. There is speculation that this once unusual behavior is spreading and has become learned. Man-whale interactions in the past have not always been as benign as the behavior just described.

From the 1845–46 season through the peak periods of 1854 through 1865, gray whales were captured and slaughtered by commercial whalers in such large numbers that the population disastrously declined almost to the point of extinction. From 15,000 whales that are estimated to have existed prior to that period of commercial exploitation, the eastern North Pacific stock of the gray whale has rebounded and expanded to about 26,000 with the protection of an international ban, treaties, and laws. In 1999 a notable increase in strandings of this stock took place, five times the number of any previous year on record. In part, the strandings can be attributed to normal whale mortality. Because the gray whale habitually swims and feeds near the coast, dead grays often strand on shore, where we can count them. We do not know what the normal death rate is for whales that die in the open ocean. Also, with increased population and protection from hunting, a few whales are now probing protected bays never before visited by this species, such as San Diego Bay, Puget Sound, and San Francisco Bay. These are waters that sustain all manner of pollutants and human activities that possibly are harmful to cetaceans. Even though, on 16 June 1994, the eastern stock of gray whales was removed from the Endangered

Species Act List of Endangered and Threatened Wildlife, it is important to continue to monitor these populations to discover the reason(s) for unprecedented die-offs or an increase in mortality. A population at or above carrying capacity of the environment can be expected to experience intervals of high mortality in poor feeding years.

The gray whale suffered from aboriginal human predation for about 20,000 years. The aconite poison fishery (see *When Did Whaling Begin?*) had the capacity to reduce the population considerably. The gray whale was probably driven from the sheltered waters of the Pacific coast and forced to migrate from its breeding areas in Baja California to the Bering Sea. With the stop of aboriginal hunting of gray whales in the nineteenth century and of European commercial whaling in the latter part of the twentieth century, the population began to increase. The gray whale has again come to inhabit waters that were closed to it since humans populated North America.

Rorquals and the Humpback Whale (Family Balaenopteridae)

Blue Whale (*Balaenoptera musculus*)   The generic name *Balaenoptera* is derived from the Latin *balaena* (= whale) and the Greek *pteron* (= fin); the specific name *musculus* is probably a parody meaning "little mouse." That means the blue whale's scientific name translates as "fin whale [like a] little mouse." (All the species of *Balaenoptera* are also referred to as finner whales.) Largest of all living animals is the blue whale. This whale can grow to more than 33 meters in length, surpassing the length of our biggest buses or tractor trailers (20–25 meters). If you were to stand beside the Boeing 737 airplane, which is 33.4 meters long, you would begin to understand how big a blue whale can be. The largest reported blue whale was a 33.58-meter female from South Georgia, in the Antarctic. The largest reliably measured was a 28.5-meter female. The heaviest weight reported by one author was 190 metric tons for a 27.6-meter female, also from South Georgia. Females are normally larger than males of like age, and blue whales found in the Southern Hemisphere are larger than those in the Northern Hemisphere.

Blue whales are equipped with 270 to 395 baleen plates per side, about 1 meter long and black in color. They are members of family Balaenopteridae, genus *Balaenoptera*, commonly known as rorquals. Not all baleen whales are rorquals. The name "rorqual" comes from the Norwegian *ror*, meaning "tube," and *hval*, meaning "whale," and refers to the extensive grooves that run from under the chin down the throat and chest to well beyond the flippers in some species. Rorquals typically have many grooves; the blue has 55 to 88, not all the same length. The longest run from chin to navel; the shortest, along the sides of the face. Feeding on some of the

A blue whale cow and calf as seen from the air. The long slender nature of the blue whale's body becomes apparent to us only in such shots. The impression that we can gain from such a photograph is much different from the early notion we formed from their dead flabby carcasses.

smallest of animals, primarily krill and sometimes copepods, the blue whale is equipped anatomically with the capacity for holding enormous amounts of food and water. The grooves permit the throat to expand tremendously while feeding, allowing the tiny animals (euphausiids) entrance to the vast mouth along with massive amounts of water, which is filtered out through the baleen. Rorquals include the fin whale, sei whale, Bryde's whale, and minke whale in addition to the blue whale, and they all share various anatomical features. Even though humpbacks have baleen, they are not included among the rorquals; instead, they make up a single species in the genus *Megaptera*. Their extremely long flippers, stout body, throat grooves, which number far fewer than the rorquals, and vocalizing repertoire distinguish them from the genus *Balaenoptera*.

As befits their name, blue whales are generally a bluish gray mottled with white or gray spots. The undersides of the flippers are white. "Sulfur bottom" is a common name given to some blue whales that develop a yellowish hue on their bellies, caused by an accumulation of cold-water algae called diatoms. Looking down on a blue whale from above, the head resembles a broad almost U-shaped arch. A single ridge extends down the middle of the raised part of the head forward of the blow-holes almost to the tip of the snout. The dorsal fin is small and located far back on the body, about a quarter of the way forward from the notch of the fluke and is normally seen only when the whale starts to dive. The body appears to be broad with two concave areas in front of the lungs, which expand and contract when the whale breathes. Notwithstanding its huge size and status as the largest living animal, the blue whale with its long slender flippers and tail stock appears slimmer in the water than the sperm whale.

Blue whales are found scattered in all oceans. Migration routes seem to be determined by the abundance of food, which blues need a lot of to sustain themselves. "Follow the food" may be the byword for rorquals, indeed, for all animals. In the North Pacific in April and May, blues move toward the poles, from northern California to the eastern Gulf of Alaska, where in the spring and summer there are large blooms of krill along the Aleutians to the Bering Sea. In the autumn and winter they head toward subtropical and tropical breeding grounds off southern California and southern Baja for mating and calving. In the North Atlantic, blues migrate to Arctic waters, but their current wintering grounds and southern limits are not specifically known, and it is speculated that they are widely dispersed throughout the warmer waters when breeding. In the past there were sightings off Long Island, New York, and Ocean City, New Jersey.

In January 1922, a blue whale made a fatal mistake and swam into the harbor of Cristobal, at the northern entrance of the Panama Canal. It swam toward the locks at Gatun Lake, where it was destroyed (in the name of shipping safety); its carcass was towed back to the Cristobal docks, where attempts were made to measure it, and was finally towed out to sea. Its remains drifted ashore again and eventually were bombed by U.S. Army planes. The second and third cervical vertebrae were retrieved from the water by English explorer Mitchell Hedges, who gave them to the British Museum (Natural History). What was unusual was the appearance of a blue whale so far south at that time of the year. Although it was possibly from a North Atlantic population, because it was January, and Northern Hemisphere populations would have been swimming south at that time, those populations do not usually cross the equator in their migration. Also, most puzzling was its size, 29.8 meters, which was considered too large for a northern migrant. Such a large size would be expected in a whale from the Southern Hemisphere. In the course of migration, did this whale cross the equator from the Northern Hemisphere to the Southern, or was it moving in the opposite direction? Or was this a challenge to the assumption that Northern Hemisphere blue whales are smaller than those in the Southern Hemisphere?

In the Southern Hemisphere, summer sightings are south of 40° south latitude, where the blues feed almost solely on one species of krill, *Euphausia superba.* In winter, the whales keep ahead of the pack ice and move northward, but few sightings have been reported that indicate locations of breeding grounds. Blue whales have been reported as far north as Madagascar, Ecuador, and Peru. Even though we know where blue whales go in the summer, the mystery remains as to their exact whereabouts in winter. In the case of the whale stranded in the Panama Canal, we do not know where it came from or where it was going.

Gulping is a favorite feeding technique of blue whales. Where there are large concentrations of krill, whales enter the area, take huge mouthfuls of water, and close their mouth forcing the water out through the baleen, trapping the krill, which then may be moved backward with the large black tongue. Blue whales may dine on as much as 4 metric tons of food a day, with peak eating hours during the early morning and evening. When blue whales were taken in the subtropics where they winter and their stomachs were examined, they had little or no food in them. It is assumed that most fast on their journey to breeding sites.

We are not yet sure how long blue whales live. To determine the age of baleen whales, we use a method that analyzes growth layers in the ear plug. If the plug is cut lengthwise it shows alternating layers, or laminae, of shed skin cells (light) and wax (dark). Investigators formerly believed that two layers were formed each year, but current thinking is that only one layer per year is laid down. Because this method was developed in 1955 and populations were low at that time, we have yet to accumulate enough data to provide a definitive answer in regard to blue whales. Nevertheless, some authors estimate a maximum age of between 30 years and 80 or 90 years.

Blues have a single calf every two to three years. Depending on how one estimates the rate at which the ear plug laminae are formed, we know that males and females become sexually mature approximately between 5 and 10 years of age, in both hemispheres. In the Northern Hemisphere, they mate in late fall and winter. In the Southern Hemisphere, they mate during the austral winter with a peak in July, so that the reproductive seasons in the two hemisphere are out of phase with each other by six months. The gestation period is estimated to be 10 to 11 months. At birth, calves are about 6 to 7 meters long, but by the time they are weaned at seven months, they have more than doubled their size, growing to about 16 meters.

Blue whales are very fast swimmers, with ranges of 5 to 33 kilometers per hour while cruising or migrating and 2.5 to 6.5 kilometers per hour when feeding. The blue whale blow varies in height from 5 to 9 meters according to how fast the whale is traveling and other activities that are taking place The blow is dense and tall, whereas those of other large whales, such as humpback and right whales, are bushy. Normally, blues blow a few times at intervals of about 20 seconds, then dive for perhaps 5 to 10 minutes. Generally, the blue whale fluke rises into the air when the whale dives.

Despite its size, the blue whale has some enemies that dare to attack it. Killer whales prey upon blues. Because blues more often than not travel alone or in twos or threes, several members of a killer whale pod working cooperatively can effectively attack and kill a large whale like a blue. Such behavior has been documented on film. However, it is humans, also great predators, who have been most responsible for reducing the blue whale populations. Early whalers seldom concentrated on catching blue whales, hampered as they were with small open boats and hand-held equipment. The whales were too large and fast and, if harpooned, sank quickly before they could be retrieved. When modern whaling techniques were developed in the late 1800s and the floating factory ship was introduced in the early 1900s, the blue whale became a favored target. From 1930 to 1931, one year, 29,410 were killed. Humans definitely contributed to the decline in blue whale populations until 1966, when the species was given protection by the International Whaling Commission. Many scientists, however, wondered why the blue whales did not recover after they were protected. Hypotheses for this involved competition from other whale species for food and the reduction of the blue whale populations to a point where they could no longer repopulate or maintain themselves, so they just gradually dwindled. A simpler reason was revealed in 1995 with the admission that the Russians had continued to take blue whales, long after the prohibition. Now, scattered populations, for example, off Monterey, California, and off Baja California, are recovering. The newest danger to blues can also be attributed to man: when ships and whales collide.

Fin Whale or Finback (*Balaenoptera physalus*), Sei Whale (*Balaenoptera borealis*)
The fin whale, *Balaenoptera physalus* (*physalus* from Greek, meaning "a kind of whale"), is another enormous rorqual. Second in size only to the blue whale, females grow up to 27 meters in the Southern Hemisphere. Sei whales, *Balaenoptera borealis* (*borealis* from Greek, meaning "northern"), also rorquals, grow to 20 meters in length. Fin whales are reputed to take first place in the category of speed, reaching more than 32 kilometers per hour in short bursts. Radio tracking of a tagged fin whale that cruised 292 kilometers in one day showed average speeds of 3.6 to 14.6 kilometers per hour.

In appearance rorquals are quite similar. They are slender, some a little heavier than others, and streamlined. They have grooves, or pleats, that run from under the chin to the ventral region, varying in distance from the navel according to the species. The fin whale has about 50 to 100 grooves, and they run to the umbilicus and somewhat beyond. The sei has fewer grooves, which vary in number in the Northern and Southern Hemispheres but commonly range from 38 to 62. These do not reach the naval. The fin whale body shape is rounder in front than the blue but not as bulky as the sei whale. Looking at a fin whale from above, the head appears triangular, or V shaped compared with the blue whale, which looks U shaped. The sei whale head in appearance is somewhere between the two others, not quite triangular and not quite U shaped but with a downward tilt at the tip of the head. The head of both sei and fin whales takes up to 20 to 25 percent of the body length.

Several characteristics help us to distinguish between the fin whale and other rorquals. The fin whale's dorsal fin is much larger than that of the blue. It is about 60 centimeters tall, whereas the blue's is but a nubbin of a fin. Sometimes called razorback, the fin whale has a prominent median ridge that runs from the dorsal fin to the flukes. The sei dorsal fin is larger than that of the fin whale and visible when the whale blows. It is fairly erect but pointed far backward. Similar to the fin whale's, it is located about two-thirds of the way between head and tail. The sei whale blow, which can reach 3 meters, is shorter than the fin whale's, which reaches 4 to 6 meters.

The coloration of the fin whale is curiously asymmetrical. It is dark gray on its head, fading to brownish black on its back and sides, to white on its ventral surface and undersides of the flukes and flippers. There is a gray white chevron on the back of some individuals; the long arms of the chevron point backward toward the flukes. The dark color on the head runs farther down on the left side than on the right. The right lower jaw in front is white, and the right front baleen plates are white or yellow white. The plates on the left side are dark blue gray, showing some streaks of white or yellow. The fibers hanging from the plates are almost all yellow white. There are 260 to 480 baleen plates, which share this strange coloring.

The sei whale is more uniformly colored than the fin. Its head is dark gray, and its belly is often lighter with light- colored scars and patches. The baleen plates, which number 318 to 402, are dark gray with pale fringes.

Fin whales are found worldwide but more often in temperate, Arctic, and Antarctic waters rather than in tropical seas. They regularly migrate in the winter to temperate waters, where they mate and calve, and in the summer to polar waters, where they feed on a variety of krill, fishes such as herring, cod, and capelin, and squids, mainly using an engulfing technique.

Fin whales are often found in groups of as few as 3 to as many as 10 to 20; however, single whales or pairs are also common. Fins rarely breach, but when they do, they leap clear of the water, then fall back into it with a great noisy splash. They can also dive to depths of at least 230 meters, deeper than sei whales.

Female fin whales are larger than males. Both reach sexual maturity at about the same age, between 6 and 12 years, and in the Northern Hemisphere at about 18.3

meters in length for females and 17.7 meters for males. Females are pregnant for about 11 to 12 months and give birth to a single calf every 2 or 3 years. Calves are weaned after 6 or 7 months when they are about 12 meters in length. Sei whales probably mate and calve in a similar fashion, depending on their seasonal migrations to keep them in favorable waters when they reproduce and feed.

Humpback Whale (*Megaptera novaeangliae*)   Without doubt, the whale that has most captured the imagination of the public is *Megaptera novaeangliae*, the humpback. It has become the poster child of the whale world. It is a species easily recognized by its huge white flippers and knobby head, a whale whose repertoire of songs has been popularized by the recording industry, a whale known for its exuberant aerial behavior, leaping out of the water, breaching, lobtailing, and flippering. Through television and film, we have welcomed this huge animal into our homes, and as a consequence of our familiarity with its appearance and behavior, the national consciousness has been raised about the need to preserve and save endangered whales.

Among the Balaenopteridae, the humpback's anatomical differences are enough to place it in a separate genus from other rorquals. *Megaptera* means "large wing"; *novaeangliae*, "New England[er]." Its most distinguishing features are the unusually large white or partly white pectoral fins, about a third of its total body length. The

The small dorsal fin of the humpback, shown in this photograph, is situated on a base of thick blubber. That, coupled with the way the whale arches its back when it is preparing to dive, gave the humpback its common name.

Another of the many diagnostic characters of the humpback is its enormous flippers. They are roughly a third the length of the whale. They function to dissipate heat that the whale generates in tropical waters, and they provide maneuverability that the humpback needs in some of its feeding behavior. (Photo obtained under NMFS permit #987)

A humpback cow and calf in the tropical Hawaiian breeding waters.

fins have many knobs or tubercles (often white) on their leading edges, as the upper and lower jaws and rostrum do. In the center of each knob on the lower jaw and head is one whisker, or vibrissa. It has been suggested that the regularly spaced knobs increase hydrodynamic lift by controlling the flow over the flippers. The great length of the flippers enhances the maneuverability needed for the humpback's unique feeding behavior. The flippers may also radiate excess body heat during humpbacks' winters in shallow tropical waters, where there is no colder or deeper water they can reach to cool off.

The throat grooves for this species are wide and number from 14 to 22, far fewer than other North Atlantic baleen whales, which normally have 38 to 100 grooves. Unlike other baleen whales, the rear margin of the tail is prominently serrated. If you look at the head from the side, it appears to be slim because of the narrow rostrum, but looking down, it appears broad and rounded. The dorsal fin is variable in

One of the ways that a humpback uses its flippers is in the behavior known as flippering, where the whale slaps one or both of its flippers onto the water. This is a lot like "lobtailing," and it may function for the same purposes, which are not yet entirely clear to us.

shape and size, but it often includes a small step or hump that is most noticeable when the whale arches its back as it begins to dive.

All humpbacks are black in coloration on their dorsal surface, but they are adorned with variations of all black, all white, and black-and-white marbling on their ventral surface. The pattern on the undersurface of the tail is also variable, ranging from black to white and combinations thereof, so that individual whales can be identified by these patterns together with the serrations on the edges of the tail. On each side of the mouth are 270 to 400 black baleen plates. Sometimes the plates farthest forward are dull white or partly white.

Of several thousand animals taken from Antarctic and Australian waters between 1949 and 1962 and measured by the biologist R. G. Chittleborough, the largest humpbacks were a 14.3-meter male and a 15.5-meter female (Chittleborough 1965: 69). Among other unpublished, but supposedly reliable, measurements from the Southern Hemisphere are records of a 17.4-meter male and a 16.2-meter female taken in the Antarctic in 1934–35 (Clapham and Mead 1999: 2).

Externally, males and females are similar in appearance except for the urogenital area. There, females have a hemispherical or crescent-shaped lobe at the end of the genital slit; this lobe is absent in males. Further, the distance between the genital slit and the anus is much greater in males than in females.

Humpbacks range throughout all the oceans of the world along the coasts and shelf waters and, during migration, will travel across deep waters. Their migrations are among the longest of any mammal and are known to reach almost 8,000 kilometers. Because they are highly migratory, their distribution changes with the season. In spring, summer, and autumn, humpbacks feed in temperate, or high-latitude, waters; in winter they migrate to lower latitudes for mating and calving in the tropical or subtropical waters where there are islands or offshore reef systems. The winter range of some may even extend to equatorial waters. The only known exception to this seasonal pattern is the population in the Arabian Sea, where they both feed and breed in the same waters. They are seldom seen in the Mediterranean Sea.

In the western North Atlantic, humpback feeding grounds span the eastern coast of the United States to western Greenland. In the eastern North Atlantic, feeding areas exist around Iceland, northern Norway, Bear Island, and Jan Mayen Island. In winter North Atlantic humpbacks from virtually all areas breed in the West Indies. Food consists of various species of schooling fishes (herring, mackerel, sand lance, sardines, anchovies, and capelin) and several genera of euphausiids (shrimplike crustaceans).

In the North Pacific, humpback feeding grounds encompass a wide arc from California to Alaska and along the Aleutian chain into the western North Pacific. The food they eat is similar to that of the North Atlantic populations. Major breeding grounds are located off Mexico, Hawaii, and Japan. Those humpbacks feeding off Alaska migrate primarily to Hawaii, whereas those from California to Washington breed inshore in Mexican waters. From the western North Pacific, humpbacks may target breeding areas in Japan.

Humpbacks from five feeding areas in Antarctic waters migrate to separate breeding grounds off Australia, Africa, Oceania, and South America. Humpbacks in the Southern Hemisphere primarily eat euphausiids.

Data from the whaling industry suggest that migration to and from the tropics is ordered according to sex and maturity of the animals. First to leave the feeding grounds for the trip to the tropics are lactating females (those secreting milk), then immature animals, mature males, "resting females," and last of all, pregnant females. The order is reversed in late winter when the whales migrate back to the feeding grounds.

Reproduction is also strongly seasonal, with ovulation and conception taking place in the winter in warm tropical waters. This is followed by a gestation period of 11 to 12 months and migration back to the cooler waters of the feeding grounds. The cycle is then completed by the return to warm waters to give birth. Sexual maturity is attained by males and females at the average age of five. Physical maturity is not reached until 8 to 12 years after sexual maturity. In the Southern Hemisphere, ovulation occurs from June to November, with a peak in late July. In the Northern Hemisphere, ovulation occurs in the alternate six-month span.

The growth rate of the humpback fetus is quite fast, exceeded only by those of the blue and fin whales. There are records of twin fetuses but no reliable records of a female nursing more than one calf. Calves, usually 4.5 meters in length, are born in early August in the Southern Hemisphere and in early February in the Northern Hemisphere. They grow swiftly, consuming at least 43 kilograms of highly nutritive milk a day and spurting to between 7.5 and 9 meters in length by the end of lactation. Nursing lasts for about a year, but some calves begin to feed independently at about six months. After giving birth, mother and calf remain intimately associated, the mother nursing and pushing her young one upward to the surface to breathe. Most calves leave their mothers sometime before the second winter, but a few remain associated with the mother for as long as two years. Humpbacks commonly give birth about every two years.

The mother and calf are often followed by another whale, a male who seems not so much interested in baby-sitting as in keeping his eye on the mother for the first opportunity to breed. The winter grounds are not only used for giving birth and nursing but are also where mating occurs. Mating is a dangerous activity, particularly for a year-old calf following its mother. As many as five or six males may attempt to surround or jockey for position in attempts to attract the female, even if she is still nursing.

The current situation with humpbacks is poor for learning much about their life expectancy, because their populations were greatly reduced and most of the largest and oldest animals were killed in whaling activities in the past century. Humpbacks are frequently found in coastal waters and are slow swimmers when migrating (ranging from 2.2 to 8.2 knots, or 7.9 to 15.1 kilometers per hour); therefore, they were often the first species chosen by whalers to harvest. In the Southern Hemisphere, more than 200,000 were harvested in the last century. Humpbacks were finally accorded protection in 1966, but because so many large ones were taken, their life expectancy is difficult to predict today. The oldest, out of many thousands taken in Australia, was reported by one researcher to be 48 years old (if the age determination technique is correct in assuming that two growth layers per year are laid down in the ear plugs).

Of all the large whales, the humpbacks are the most vigorous, acrobatic, and diverse in their behavior. Their activities might be loosely divided into three groups—feeding behavior, courtship, and mating behavior—but they also exhibit behaviors that could take place anywhere throughout the year in various situations, both by single whales or by those in groups.

There are several types of feeding behavior among humpbacks. When there are large patches of krill near the surface of the water, these whales may use a lunge-feeding technique, rising swiftly in the water column, mouths agape to engulf their prey. Another spectacular method, bubble netting, is unique to humpbacks and is used by one or several of them working together to corral or trap prey, usually schooling fishes such as herring. The whales produce a cloud, or curtain of bubbles, through the blowholes. The bubbles rise to form a ring, concentrating fishes in the center and trapping them between the surface of the water and the whales' mouths below. All the cooperating whales then have to do is swim vertically upward through the circle with their mouths open and swallow the trapped fishes. What an astonishing sight it is to see these enormous creatures capturing prey: heads surfacing, huge mouths open, throat grooves bulging with water and fish. There is also evidence that humpbacks feed on the sea floor.

Perhaps the most distinctive and well known of the various kinds of behavior associated with courtship and mating is the singing of the humpback. *Songs of the Humpback Whale*, a long- playing record made in 1976, based on the research of biologist Roger Payne and featuring the eerie, mysterious vocalizations, captured the public's imagination and inspired concern about the plight of endangered whales. Music lovers became conservationists; the songs of the humpback touched an emotional chord that led to new interest in all aspects of whale behavior. Songs are sung only by solitary mature males almost entirely on the breeding grounds, primarily to attract females. The periods of singing are long, often lasting hours or even days. The individual songs consist of several themes sung in a generally invariant order, with one song lasting a few minutes or as much as half an hour. All the males in a certain population sing the same song; the differences or similarities of songs have been used to differentiate populations. The song changes progressively, both over the course of a breeding season and over a period of years, but somehow all singers keep up with the changes. The exact mechanism of these song changes is unknown, but presumably sexual selection (females selecting males whose songs they like) plays a key role.

Singing always occurs at the winter breeding grounds, will sometimes be heard on migration, and occasionally has been recorded in the summer and fall at feeding grounds. Other than songs, humpbacks produce sounds while migrating that include groans, cries, chirps, shrieks, and clicks. On the feeding grounds, the sounds recorded have been grunts, yelps, moans, and trumpeting through their blowholes.

Other kinds of humpback behavior that occur together with other whales or alone and at various times of the year include breaching, or leaping out of the water, perhaps to signal their position or from excitement or the need to play. Flippering, where the whale slaps one or both pectoral fins onto the water, and lobtailing, where the whale slaps the tail forcefully down onto the water, are actions not well understood. It is believed this behavior may be aggressive in nature.

TOOTHED WHALES (SUBORDER ODONTOCETI)

Sperm Whales (Family Physeteridae)

Sperm Whale (*Physeter catodon*)   The sperm whale is a magnificent animal of enormous proportions, a creature of extremes. Of all the toothed whales, it is the largest. This is a sexually dimorphic species, with males larger than females. Males may grow to a maximum of more than 18 meters and females to about 12 meters. Determining the weight of a whale is a difficult problem. It has been accomplished with a crane and scales. Think of the size they have to be to lift and weigh such a beast. A whale weighs 13.8 percent more when it is whole than when cut up. If cut into pieces, all the blood, spermaceti, and other fluids are lost from the body. For many years, the heaviest sperm whale recorded was a 16-meter male that weighed 39 metric tons (Boschma 1938: 152). Another more recent report cites an 18.1-meter male weighing 57.1 metric tons and an 11-meter female at 24 metric tons (Lockyer 1979: 273).

The sperm whale is easy to identify with its distinctive head—huge and shaped like a box. The head alone is a quarter to a third of the length of the entire body, which is another aspect of the sperm whale that is extreme. The brain is the heaviest in the animal kingdom but only 2 percent of the total weight of the whale. When 16 male brains were weighed, their individual weights ranged from 6.4 to 9.2 kilograms. The upper portion of the cranium holds the long barrel-shaped spermaceti organ (see *What Is Spermaceti?*), and in males, particularly, it projects forward over a narrow lower jaw that can hardly be seen but holds about 17 to 30 pairs of teeth. In contrast with the dark gray body of the sperm whale are the white upper lips and lower jaw rim. It is rare for a sperm whale to be pale or entirely white, as Moby Dick was in Melville's classic tale. In 1957, however, a Japanese catcher boat did kill a white sperm whale (the pink iris and red pupils of a true albino were not noted). Photographs of the 10.5-meter male show an extraordinary white whale. There are cases of complete albinism. In 1952 two unusually light albinos with pinkish white eyes were caught. More recently Soviet fleets in 1966 in the south-

A sperm whale swimming. The body shape of the sperm whale is truly unique, with its large squared-off head and short stubby flippers. The skin is thrown into rigid corrugated folds, which are diagnostic of this species.

ern portion of the Gulf of Alaska caught two, one a female with a very small area of gray area under the fins. We present a photograph of a white sperm whale calf in Chapter 3 (see *How Have Whales Been Discussed in Folktales and the Arts?*). Light gray sperm whales are also quite rare, but there is a recorded case of a light gray male over 15 meters long caught in the Antarctic in December 1950. Without doubt, however, most sperm whales that range the oceans are of the darker colors: from gray or dark bluish gray to gray brown and black brown, with light spots around the umbilical region.

The throat has 2 to 10 grooves, and there is but one blowhole, on the left side of the head toward the front, which causes the blow to be bushy and angled sharply toward the left, rising to some 4.5 meters in height. The dorsal fin is represented by a rounded or triangular hump on the whale's back about two-thirds of the distance between the tip of the snout and the tail. The dorsal fin is followed by a series of knucklelike indentations, most visible when the whale arches its tail and dives.

In regard to diving skills, the sperm whale takes a back seat only to the beaked whales. The sperm whale is a very deep diver and can stay down for more than 90 minutes. Maximum diving depths include one extreme of 2,250 meters, recorded by passive acoustic tracking, but more often dives are less than 1,000 meters. Most of the time sperm whales swim slowly and calmly near the surface of the water at about 7.5 kilometers per hour, but in a chase or when alarmed, they can accelerate briefly to 30 kilometers per hour. At the surface they blow five or six times a minute.

Sperm whales live in all the deep oceans of the world. They roam from the equator to higher latitudes near the edges of the polar ice packs. Males range farther from the equator than females do. One of the few areas they do not inhabit is the Black Sea. Sperm whale populations have been estimated at about 3 million prior to 1864, when modern whaling began with the use of a harpoon gun mounted on small catcher boats. Present-day populations are now estimated to be fewer than 2 million. Mortality rates for males are reported to be somewhat higher than those for females. The maximum life span is estimated at 60 to 70 years using tooth-layer counts, but it is difficult to calculate age in older animals, because the layers are not easily discernible.

Sperm whales have few enemies, probably because of their large size. Nevertheless, a massacre of major proportions by killer whales occurred only recently, in 1997, about 110 kilometers off the coast of California (see *What Do Whales Eat?*). Although killer whales are reported to prey on newborn calves of any whale species, this attack was quite different. Nine sperm whales, thought to be adult females, tried to protect themselves by gathering into a wheel or rosette formation. The sperm whales positioned themselves with their heads toward the center and their bodies and tails extending outward, like the spokes of wheel, for defense. The attacking killer whales were also females, some accompanied by calves. At the onset, 3 or 4 killer whales circled round the sperm whales and tried to pull off chunks of flesh, but eventually the number of attackers grew to about 40 or 50. Robert L. Pitman and Susan J. Chivers, scientists with the National Marine Fisheries Service, reported the attack from the *RS David Starr Jordan*, a research ship of the National Oceanographic and Atmospheric Administration (NOAA). Their eyewitness account reports that, even though the sperm whales were much larger than the killer whales, they did not appear to defend themselves adequately in the face of the remarkable charges and cooperative hunting strategies taken by the killer whale pack. When one of the sperm whales unexpectedly left the rosette formation, it immediately became vulnerable to attack and was severely wounded by killer whales on both sides, even though two other sperm whales detached from their formation to come to its aid. The story of this attack is told in *Natural History* magazine and brings to light new information on sperm whale behavior, challenging former ideas that sperm whales, by virtue of their size and deep-diving abilities, are spared from predation by killer whales (Pitman and Chivers 1999). We have no answers as yet but only confirm that this is another area of whale life about which we know little.

Sperm whale blubber has also been found in the stomach of blue and Portuguese sharks, but this may be just the result of scavenging. Historically, man has been the sperm whale's worst enemy. From 1833 to 1849, 100,000 barrels of sperm oil were imported by the United States, which represents an annual kill of 5,000 sperm whales. In 1985 sperm whaling was banned by the International Whaling Commission (IWC).

We have not yet observed sperm whales feeding, so we have to look at indirect evidence. From the examination of stomach contents of dead whales, we know that the chief food of sperm whales is squid. This includes the very largest squids in the depths of the ocean, giants such as *Architeuthis* up to 30 meters long, moderate-sized squids, and occasionally large bottom-dwelling octopuses such as the giant *Octopus dofleini*. A variety of fishes, medium sized to large, such as salmon, skates, and rockfish, are also taken from time to time, as well as nonfood objects such as rocks, sand, shells, a boot, and one report of a human corpse. The amount of food they eat every day is estimated at 3 to 3.5 percent of their body weight.

Various scenarios have been suggested for the way sperm whales attract or find food, but we should emphasize that all of these are hypothetical. Some ideas revolve around the fact that many of the squids composing the whale's diet have luminescent organs, making them visible to an approaching whale. In some scenarios, the whale could swim with its jaw lowered while randomly searching. Because sperm whales have been captured with jaws that are sometimes fractured or deformed even though the rest of the animal is healthy, it is suggested that they plow up the mud with their lower jaws to obtain squid and sometimes injure themselves doing so. Echolocation might be used to scan for prey, or the whales might remain motionless with their jaw lowered, hoping that prey will be attracted to their white lips. This could be particularly true if the jaws were smeared with the luminescent mucus from squid they had eaten. Another speculation suggests that sperm whales can stun and immobilize prey by giving off extremely strong pulses of ultrasound.

Sperm whales have signature sounds. Heard through the receiver of a ship's sonar, schools of sperm whales continually emit clicks, which are sounds composed of 1 to 9 pulses each, 1 to 2 milliseconds long, with intervals of 2 to 4 milliseconds. Individual whales can be identified from the unique rhythm and spacing of pulses. These sounds are probably used for communication and echolocation, to find food, and to track the location of other whales, which may be some miles away.

Studies have been made of the reproduction of the sperm whale in different geographic areas all over the world. In general, females ovulate and often become pregnant for the first time between the ages of 7 and 13 years, when they are from 8.3 to 9.2 meters in length. It takes longer for males to reach sexual maturity, 18 to 21 years, when they are about 11 to 12 meters in length. Sperm whales form breeding schools and bachelor schools. Breeding schools include females of all ages and immature males. Bachelor schools are composed of pubescent and sexually mature males. Many of the largest males, however, remain solitary or sometimes gather in groups of no more than six animals. In mating season, these larger males attach themselves to breeding schools. Mating can occur from late winter through early summer. Males bearing tooth scars on their heads seem to indicate some fighting for the favor of females or for advantage among the bulls. In copulation it is thought that males and females lie horizontally, belly to belly, one above the other. The pregnancy, or gestation period, lasts 14 to 15 months, with females giving birth to one calf every 3 to 6 years. The young calf suckles mother's milk for two years, adding some solid food after one year. The adult female then goes through a resting period for another .75 to 2.75 years before she is ready for breeding again.

Unfortunately we cannot see sperm whales in captivity. They are just too big to live in oceanaria. Your best opportunity is to book an excursion on a whale-watching vessel that has had previous sightings of the species that interest you.

White Whales (Family Monodontidae)

White Whale or Beluga (*Delphinapterus leucas*)   The white whale is aptly named. Its taxonomic name derives from the Greek *delphinos*, meaning "dolphin"; *a*, "without"; *pteron*, "fin"; and *leukos*, "white." White whales are exceptional. Unlike many other whales, this species lacks a dorsal fin; instead, there is a ridge (sometimes dark) that runs along the spine just past the midpoint of the back. The lack of a dorsal fin may be a modification for life in the Arctic, because it also occurs in the bowhead and narwhal, other occupants of those icy seas. This whale's most conspicuous feature, however, is the creamy white color of the adult. Calf colors transition from a slate gray, possibly with a brownish pink tinge in the newborn, to gray, light gray, bluish white, and, finally at five or six years of age, white. These are sturdy, compact, torpedo-shaped animals with tapered ends.

The size of belugas differs according to the geographic regions they inhabit. The smallest are found in the White Sea and Hudson Bay; the largest, off Greenland and in the Sea of Okhotsk. Even within a particular geographic area, a wide range of size can occur. Length extremes from the Gulf of Saint Lawrence and Cumberland Sound populations range from 139 centimeters for the smallest newborn to 447 centimeters for the largest adult. Males in all regions grow to be about 30 to 50

A mixed school of young and adult belugas in shallow water in the Canadian Arctic. Belugas are a mottled dark gray when young and do not become completely white until later in life.

centimeters longer than females. Maximum length was reported to be 667 centimeters from the Sea of Okhotsk, but this was probably an exaggeration. A more reasonable maximum length is 599 centimeters.

Even the texture of the beluga skin is different from that of other whales. Most whales have plasticlike skin; white whale skin has been likened to the texture of putty. Other distinctive features include a small rounded head with a protuberance (known as the melon), a short broad beak, a short neck, and small flippers with upturned edges found on the male at physical maturity. The female at physical maturity has flippers that are flat. Unlike most other whales, which have fused cervical vertebrae, the beluga's cervical vertebrae are free. We know that belugas have more flexibility of the neck than other whales and can move their head up and down and from side to side, but we do not know whether the unfused vertebrae are important or necessary to this ability. The mouth of the beluga also has some flexibility that appears to endow it with greater facial expression than what other whales have, possibly an adaptation for sucking food, especially for the young, or perhaps a means of communication.

One of the problems in the beluga habitat is the extreme tides. Sometimes the beluga makes a mistake and is left high and dry until the tide returns.

Watching belugas swim, particularly in large herds, is an exciting experience. They are extremely supple and highly mobile; they rotate their fins, turn, twist and roll, and often swim upside down. In captivity they have been seen swimming backward using their fins in a sculling movement. Belugas normally cruise at about 3 to 9 kilometers per hour, but if they are chased they can move as fast as 22 kilometers per hour (without young) for intervals of 15 to 20 minutes.

These are social whales, gathering together in pods of only a few animals or in concentrations of as many as thousands to feed or migrate. Huge herds of perhaps 10,000 individuals can cover the surface of the water as far as one can see. They are also the noisiest of whales, with such a huge repertoire of sounds that early whalers called them sea canaries. These gregarious whales deliver a cacophony of sounds: clicks, squeaks, croaks, trills, chirps, squeals, and whistles, which can be heard underwater and in the air. Their sounds have been compared to a bull bellowing or pigs grunting.

Belugas mate in the spring, with some variability. After a pregnancy (gestation period) of 14 to 14.5 months, a dark, slate gray calf is born, on average 1.6 meters in length. Because of their color, newborns are sometimes misidentified as young narwhals. As the calves grow older, their color transitions from the dark to lighter gray to bluish gray and finally to white. The pure white color can be used as one indicator of sexual maturity in females and in most males. By the age of 10, when

Another feature of the Arctic is the presence of wandering polar bears. If a beluga is stranded by the tide, it is helpless when faced with a hungry polar bear.

both sexes have reached physical maturity, they are completely white. Birth takes place in the summer. Newborn calves swim closely with their mothers and drink only milk during the first year. In the second year, calves still suckle but now are able to supplement mother's milk with crabs and worms suctioned off the bottom of the estuary or river they occupy. By this time, the female is again pregnant. On average it takes 36 months to complete the reproductive cycle. Estimates from growth layers on the teeth (possibly two per year) indicate that belugas live to about 30 to 35 years.

Belugas have a circumpolar distribution, living in Arctic and sub-Arctic waters. They are common in the North Atlantic and the Canadian Arctic and along the northern and Siberian coasts of Russia. Herds are also found in Alaskan waters. Belugas winter in Hudson Bay, the estuaries of the Saint Lawrence River, the Cumberland Sound, and near Greenland in the region of Disko Bay. They are found off the north coast of Norway and in the Sea of Okhotsk in the Pacific. Belugas inhabit the islands of the eastern Arctic and the Gulf of Saint Lawrence. They occasionally travel south to the Bay of Fundy, and sightings of belugas have occurred as far south as New Jersey, but these are rare. Most white whale herds congregate near the shore rather than in the open ocean. Some herds, however, are found in deep water. Belugas are proficient at diving to depths of at least 647 meters and have the ability to remain underwater for as long as 15 minutes.

In the summer, many belugas migrate into warm river estuaries and coastal waters. We are not completely certain why they do this, but it is believed they do so to nurture the newborn and conserve energy. They adapt readily to shallow-water areas that provide a safe haven, free of predatory killer whales, walruses, or polar bears, and enough diversity of food to satisfy the young and the adults. Belugas eat a great variety of fishes, whatever is most abundant and available: herring, capelin, cod, flatfish, whitefish, and salmon. They also eat bottom organisms: octopuses and squids, crustaceans, worms, and mollusks. Records show that the beluga feeds on about 50 species in the Gulf of Saint Lawrence, and about 100 in the northern Russian seas. From studies of captive belugas, a wild adult female of 600 to 700 kilograms was calculated to eat around 4,900 kilograms of fish per year. Having occupied the bays and rivers all summer, the herds return to the ice packs in winter.

Historically, belugas in the Gulf of Saint Lawrence have been overhunted since the beginning of the last century, at least until commercial whaling for them ceased. Formerly, belugas were taken for leather for industrial belting. In the early part of the century, about 5,000 of them existed. By the time commercial whaling of them ended in the 1950s, the count was down to about 1,200. Today fewer than 500 belugas exist. The declining birth rate and premature deaths are attributed, at least in part by some scientists, to pollution from organochlorines produced by industrial activity and hydroelectric plants along the Saint Lawrence River and its tributaries on both the Canadian and United States shores. Today necropsies of many dead belugas that wash up onshore show high levels of chemical pollutants in their tissues. Diseases found in some dead whales include bronchial pneumonia, hepatitis, perforated gastric ulcers, pulmonary abscesses, and a case of bladder cancer. A large number have died from blood poisoning because their immune systems failed, triggered by the ingestion of toxic substances. Also believed by some to have a negative impact on the beluga are the many whale-watching vessels and increased boat activity in the Saint Lawrence and Saguenay Rivers.

Indeed, there may be multiple reasons for the decline in population, but the threat to the beluga's existence continues. The ability of the beluga to carry on in spite of the factors indicated above is a testament to the unbelievable hardiness of the species.

Beaked Whales (Family Ziphiidae)

Northern Bottlenose Whale (*Hyperoodon ampullatus*)   The northern bottlenose whale is one of two species in the genus *Hyperoodon*, belonging to family Ziphiidae, the beaked whales. The generic name comes from the Greek *hyperoon*, "upper story or room," and *odon*, "tooth." It refers to the former misconception that the palate is armed with many teeth. Actually, small palatal papillae were mistaken for teeth. The species name, *ampullatus*, means "bottle," referring to the similarity of the nose to a bottle. Of all the beaked whales, the northern bottlenose is the best known, mainly because of its interest to the commercial fisheries beginning in the late nineteenth century and continuing to the 1970s.

The English common name is "northern bottlenose whale." The terms *bottlenosed* and *bottlenose* are used interchangeably, but recently the tendency is to use the latter. Some common names in other languages are *anarnak* (Greenland), *andehval* and *bottelnosen* (Norwegian), *naebhaval* (Norwegian and Danish), *Entenwall* (German), and *döglingur* (Icelandic).

With a beak that makes it look like a dolphin, the northern bottlenose head is distinguished by a very large bulging forehead, or melon. The slope of the forehead is extremely steep and, along this surface, becomes flat like the end of a barrel, particularly in mature males. In general the head is much larger in males than in females. This whale may look like a dolphin but is much larger, the males growing to as long as 9.8 meters and the females to 8.7 meters.

The dorsal fin of this species is subtriangular and of moderate size, at least 30 centimeters tall, located two-thirds of the distance between snout and tail; the flippers are small with a relatively long forearm and short phalangeal elements. The flukes are broad without a median notch, as is common for all ziphiids, or beaked whales. As in all beaked whales, two short V-shaped throat grooves occur. The blowhole is wide compared with those of other whales and is located behind the bulbous forehead. The blow may rise 2 meters and is bushy.

Normally, only two teeth occur at the tip of the lower jaw, one on each side. These erupt through the gums in adult males but are unerupted in females and young. Sometimes another pair of functional teeth develops and erupts, but when this happens the ones at the tip of the jaw are always the largest. Ten to 20 vestigial teeth are also present in some individuals. These can develop in both upper and lower jaws, reaching a length of about 5 centimeters in males but a shorter length in females.

The skulls of both northern and southern bottlenose whales can be readily differentiated from other beaked whales, because they have enlarged crests on the facial surface just anterior to the nasal passages (maxillary crests). These are more

A northern bottlenose whale just surfacing to blow. The prominent beak and forehead are clearly visible in this photo. Less clear is the trailing margin of the flukes, which, in beaked whales, do not have a central notch.

developed in males of the northern bottlenose. We do not know their function, but some researchers hypothesize that they act to modify the sounds emitted by the whale.

Young northern bottlenose whales vary in color between black and chocolate brown on the dorsal surface. Some scientists think the chocolate brown may be the result of a diatom coating rather than their true skin color. Their belly is grayish white. Whether juveniles are darker dorsally and lighter ventrally or they are uniformly colored, developing countershading only later in life, is a contentious subject. Adults are darker dorsally and lighter ventrally. White or yellow white spots can appear on the sides or bellies and increase with age. A white band may occur on the neck of older females, causing them to have been known as *ringfiskar* (ring fish) by Norwegian whalers during the heyday of whaling. The older males are recognized by a white forehead. This white area always has well-defined borders and may extend as an irregular patch as far back as the eyes.

Some differences of opinion exist among researchers with regard to numbers of different types of vertebrae in northern and southern bottlenose whales, but the total vertebral count for both species remains at 46. The genus *Hyperoodon* is unique among the beaked whales in having all the cervical vertebrae fused into one mass.

Little is known about the evolution of ziphiids, or beaked whales. They first appeared in the early Miocene. Only a few fossils have been found, consisting of water-worn rostra, or beaks, but they reveal very little information of the past history of this family of whales. We actually know of no fossils that can be assigned to genus *Hyperoodon*. On the basis of geological evidence, *H. ampullatus* and *H. planifrons* (the southern bottlenose whale) may have diverged as recently as 15,000 years ago. Differences between the two species in the way the maxillary crests of the skull develop led one researcher to argue that the separation of the two species took place considerably more than 15,000 years ago (Davies 1963: 114).

Bottlenose whales are largely deep-water animals seldom found in waters less than a kilometer deep. The northern bottlenose is found in the North Atlantic from Nova Scotia to about 70° north latitude in the Davis Strait, to 77° north along the east coast of Greenland, and from England to the west coast of Spitsbergen. They have been caught off Iceland, the west coast of Spitsbergen, Jan Mayen Island, and the Faeroes. The easternmost record of their normal distribution seems to be west of Bear Island at 17° east longitude.

Northern bottlenose whales have never been caught in the North Sea but have been reported stranded on the coasts of Belgium, Denmark, France, England, and possibly Spain. We have a few documented records of these whales in the Mediterranean Sea. In the western Atlantic, the two main centers of their distribution are "the Gully" just east of Sable Island, Nova Scotia, and the Davis Strait, off northern Labrador, with stranding records as far south as Rhode Island and recent sightings off the coast of North Carolina. The northernmost records for the western Atlantic are from nineteenth-century whalers.

Not much is known of northern bottlenose distribution outside the whaling season. Some evidence exists that they follow a pattern of migrating to temperate latitudes around the British Isles in the winter and to the Arctic or sub-Arctic in the summer. Some, however, may stay in high latitudes during the winter.

One beneficial effect of the commercial fisheries was that they were required to take biological data for animals caught after 1930. This provided us with records that gave us some understanding of the reproductive timetable and behavior of the northern bottlenose. From the records of 5,224 animals, we have learned that the mean length at sexual maturity is 690 centimeters for females and 750 centimeters for males. The mean age for attainment of sexual maturity is 11 for females and 7 to 11 for males. Bottlenose whales live to at least 37 years of age and probably longer than that.

Northern bottlenose whales have a single calf, about 350 centimeters in length, every two to three years. Pregnancies last about 12 months, and the calves nurse for at least one year, judging by the presence of milk in the stomach of a one-year-old calf. Lactation may be prolonged in this species. One biologist reported seeing females accompanied simultaneously by a newborn and a yearling (Ohlin 1893: 8).

Northern bottlenose whales are social animals, usually found in groups of up to four individuals. In spring and summer, as many as 10 have been observed together but rarely more than that number. A recorded observation of three groups of more than 20 whales each off the west coast of the British Isles was unusual (Evans 1976: 13).

Northern bottlenose whales are extremely curious; they will often investigate a boat drifting or lying still in the water and will swim about it until their curiosity is satisfied. This inclination was exploited by whalers, who would arrive at the bottlenose hunting grounds and wait quietly until the whales came to investigate their ship. The whales were then easily taken. Harpoon guns were mounted on both the bow and stern to minimize the need for maneuvering for a clear shot.

Another behavioral trait that makes this whale so vulnerable to hunters is its instinct to stay with any wounded companion until that whale dies. Wounded animals sometimes were left in the water by the hunters for as long as possible in hopes that the pod would remain together and accessible, giving the whalers a chance to capture all the animals.

In early spring, at the beginning of the whaling season, the whalers found mixed herds, including newborns. In the fall, at the end of the season, they found solitary old males. The mixed herds are thought to be mating schools, and when mating is completed, the older males separate and can be found singly, similar to sperm whales.

Northern bottlenose whales are known to vocalize. One group recorded off the Nova Scotian coast whistled, chirped, and clicked with a repetition rate of clicks as high as 82 per second.

Whales of this species dive deep for as long as one or two hours at a time. They blow every 30 to 40 seconds while at the surface. One author wrote of harpooned animals remaining submerged for two hours and coming to the surface "as fresh as if they had never been away" (Gray 1882: 727) (see *How Do Whales and Dolphins Breathe?*). Harpooned whales tend to surface after dives in approximately the same place as they dove. Apparently, they do not travel very much horizontally while submerged. Normally, the flukes of the northern bottlenose do not lift clear of the water at the beginning of a dive, but if startled by the presence of vessels, the whale may show its flukes. These whales have also been observed to lobtail and breach. Recently Sascha Hooker and Robin Baird attached time-depth recorders to a northern bottlenose whale with a suction cup and got records of dives up to 1,453 meters and up to 70 minutes in duration (Hooker and Baird 1999).

Northern bottlenose whales primarily eat squid. They have a main stomach (all ziphiids lack a forestomach) with nine pyloric compartments. We can tell from examining the stomach contents of captured or stranded whales from the North Atlantic that herring is also taken, and even a starfish was found, the latter leading one researcher to suggest that these whales may eat on the bottom. In whales taken off Iceland and Labrador, squid formed the bulk of the diet, but cusk, lumpsucker, and redfish were found as well in the whales taken off Iceland, and halibut, redfish, rabbitfish, spiny dogfish, ling, and skate were found in the ones taken off Labrador.

Some parasites infect northern bottlenose whales. These animals are subject to whale lice infestations on their fins and body. Stalked barnacles sometimes adhere to the erupted teeth of adult males, and endoparasites have been found in the digestive system.

The northern bottlenose whale also has enemies. Killer whales, like humans, take advantage of its habit of remaining with wounded and dying companions. The most efficient predator of all, however, has been man, capturing and killing these animals by the thousands, first using only a few vessels and finally expanding into large commercial fisheries. The story of the development of this particular fishery is more or less the story of what has happened to many of our whale species in terms of their exploitation.

The first ship to take *Hyperoodon* was the *Chieftain* of Kirkcaldy, Scotland, which caught 28 in Frobisher Bay in 1852. In 1883, Captain Gray returned to Dundee with 200 northern bottlenose whales. The Norwegians soon took up the fishery and by 1891 employed 70 ships, taking 3,000 whales. From about that time, the fishery became dominated by the Norwegians.

The rapid development of the fishery was due to the discovery that bottlenose whale oil contains spermaceti. A fat body in the forehead between the two maxillary crests is exceptionally rich in this substance. After dissecting a number of beaked whales, one of the coauthors (Mead) became convinced that the fat body, though it contains spermaceti, is homologous to the melon of other toothed whales and to the "junk," not the spermaceti organ, of sperm whales (also Tressler 1923: 642).

The northern bottlenose was usually fished from schooners and cutters of 30 to 50 metric tons with four to six harpoon guns on the bow and stern and one in every boat. The harpoons did not carry explosive heads. If a harpooner missed the whale, the report of the gun and the sound of the harpoon striking the water did not seem to bother the animal, so the harpooner had another chance. When a whale was hit, it dove with incredible speed. One observer reported seeing 500 fathoms of line run out in less than two minutes. If the whale received a slight wound, it would take out 100 to 200 fathoms of line, go backward and forward at that depth, and surface in half an hour to an hour. When the whale surfaced, another boat fired a second harpoon. The whale dove again for a brief time, surfaced, and was killed by lancing. An adult male yields 100 to 150 kilograms of spermaceti.

The flesh of the northern bottlenose whale is quite palatable, but the oil is strongly purgative. Greenlanders are familiar with this property, and their name for this whale, *anarnak,* means "that which purges."

Even though the Norwegians are known to have started "modern whaling" practices in the mid-1800s, their fishery can be divided into two periods, an early period (1882 to the 1920s) and a modern period (the 1930s to 1973). During the early period, 60,000 northern bottlenose whales were captured; the annual maximum was 2,864, in 1896. These figures could be expanded by allowing a loss rate of 20 percent, giving a figure of 72,000 for the total kill for that period. When the modern period began, a government license was required, and the whalers were obliged to record and submit biological data on their catch. A total of 5,043 bottlenose whales were taken during the period 1938–69. The fishery ceased in 1973, when only three bottlenose whales were caught. The reasons given for the fisheries' closure were largely economic, but there was ample evidence that the population was depleted.

This is not the end of the story, even though the northern bottlenose whale is not now being hunted for commercial use. Beaked whales are incidentally taken in the pelagic fisheries, and it is possible that a few of these have been bottlenose whales. Also, recent concern has arisen over the possible effects of oil exploration and other activities in "the Gully," the submarine canyon that forms a habitat for one population of bottlenose whales.

Southern Bottlenose Whale (*Hyperoodon planifrons*)   Far less is known about the biology and behavior of the southern bottlenose than the northern bottlenose, because it never became a whale of interest to the commercial fisheries. Therefore, we have no population estimate or even rough figures on relative abundance of this species.

The southern bottlenose differs from the northern bottlenose in skull features. The maxillary crests of the southern are smaller and lower than those of the northern, and they extend backward and flow together with the occipital crests.

Common names for *Hyperoodon planifrons* in languages other than English are: *ballena nariz de botella* (Spanish), *gran calderon* (Spanish–Latin American usage), *minami tokkuri kujira* (Japanese), and *ploskolobye butylkonos* ("flat-headed bottlenose" in Russian).

The southern bottlenose is found in noncoastal waters, as the northern bottlenose is, but the southern is circumpolar in distribution in the Southern Hemisphere. There are records from the Antarctic to about 30° south latitude. Records from northwestern Australia and from Brazil indicate that *H. planifrons* occurs in warm temperate waters. It has also been recorded in South America off the coast of Chile, south and east to Tierra del Fuego and the Falkland Islands, and up the east coast to as far as southern Brazil. Stranding data have come from South Africa, Australia, New Zealand, and Antarctica. But there are problems with sighting data; some sightings are considered possible misidentifications. A ziphiid is difficult to distinguish at sea. Some species thought to be *planifrons* might be another beaked whale, *Berardius arnuxii* (Arnoux's beaked whale).

We have very few data on the appearance of the southern bottlenose whale. They are rarely seen, but those adults and juveniles for which there are data have been very similar to the northern bottlenose. One researcher reported a 650-centimeter animal seen at 48° south latitude as having the "dull yellow color that characterizes the animal in old age" (Wilson 1907: 6). Another report gives this description of two animals: "They were evidently old males, about 30 feet in length, and of a light brown color with white spots" (Lillie 1915: 118). Published photographs of a 698-centimeter female show the body to have been uniformly pigmented above and below. The dorsal surface of the head seemed to have a lighter color than the rest of the body. There were numerous white spots on the ventral surface near the rear end of the whale, and to some scientists these represented scars inflicted by squid. However, the white scars on beaked whales can be the result of a number of factors and should not be attributed to a single cause.

Details of southern bottlenose reproduction are not yet known. A 245-centimeter calf that still had fetal folds has been reported.

We have only one record that indicates the diet of the southern bottlenose whale: The stomach contents of a stranded animal consisted of squid beaks.

Oceanic Dolphins (Family Delphinidae)

Killer Whale (*Orcinus orca*)   The killer whale is one of the best known and most easily identified of all the whales and dolphins. Its bold, beautiful, black-and-white coloration is striking. The black-and-white image of the killer whale leaping from the water has been emblazoned on T-shirts, adorns posters, and is featured in advertisements. Millions of people have visited zoos and oceanaria throughout the world where killer whales have been kept captive. Some people question the ethics

One of the characteristics of the killer whale is the growth of an enormous dorsal fin in adult males. The dorsal fin shape serves as an identifying characteristic in studies of killer whale populations. The individual in the foreground is an adult male; the others are either females or juvenile males.

of keeping such an animal in captivity, and, indeed, their live capture in inshore waters of Washington and British Columbia has been banned since the 1970s. Observing killer whales in captivity, however, gives scientists the opportunity to study the animal more closely and the public the opportunity to view and appreciate an extraordinary animal that lives in the sea. Its name alone inspires our curiosity about its habits. Early on, whalers gave it the name "whale killer," because it preyed not only on fishes, penguins, and seals but on large baleen whales and dolphins as well. Somehow, through the years the name became reversed as "killer whale," perhaps a testament to the ferocity and cunning of its attacks on larger beasts. Its scientific name, *Orcinus orca*, is from the Latin *orca*, meaning "a kind of whale." Its generic name, *Orcinus*, could be a diminutive of *orca*, or it could be derived from *orcynus*, meaning "a kind of tunny." The vernacular name "orca" is often used instead of "killer whale." A variety of common names abound in other countries: *kosatka* in Russia; *shachi* in Japan; *epaulard* in France; *sverdifscur* or *huyding* in Iceland; *starhynning* or *staurhval* in Norway; *spachuggare* in Sweden; *zvarrdwalvis* in Holland; and "swordfish" in Newfoundland.

Even though it is small compared with baleen whales, the killer whale is the largest species of the family Delphinidae, the oceanic dolphins. Its shiny black dorsal surface contrasts with the white ventral surface, which extends from the tip of the lower jaw backward and partway onto its flanks, ending beyond the anus. Posterior to the dorsal fin is a light gray area in the shape of a cape or saddle. Individual whales may differ in the shape of the cape, and variations in the patterns of pigmentation may indicate differences in the geographic areas where the whales are found. The undersides of the flukes are white. The head of the killer whale is rounded with a short fat beak. One feature that makes it easy to identify the killer whale in the ocean is the oval white patch found above and behind the eye. The flippers are oval in shape and in males reach up to nearly 175 centimeters in length. Another distinctive feature in adult males is the exceedingly long, erect, triangular dorsal fin ranging from 1.0 to 1.8 meters in height. Females have a shorter dorsal fin, less than 0.7 meters in height, which is falcate (curved like that of a dolphin) rather than triangular. Killer whales are sexually dimorphic in other ways too. Females are smaller than males, reaching a length of 7.7 meters. Males may grow to be 9.0 meters long.

Killer whales are widely distributed; they are found in all the seas and oceans of the world but in their greatest numbers in the colder waters of the Northern and Southern Hemispheres, where one of their favorite prey, seals, abound. They are found as far north as the Chukchi and Beaufort Seas in the Arctic Ocean. In the northeastern Pacific, they roam the eastern Bering Sea, near the Aleutian Islands, south of the Alaska Peninsula, and the waters of Kodiak Island. They are abundant in coastal areas and the higher latitudes and may enter bays, estuaries, and the mouth of rivers. They occur in the offshore tropical waters of the Pacific, the Gulf of Panama, and the Galapagos Islands. In the western Pacific, they occur along the coast of the Bering Sea, the Sea of Okhotsk, and the Sea of Japan. In the North Atlantic they are widely distributed, from Greenland and Iceland to the waters off Norway, the United Kingdom and Ireland, Denmark, Holland, Belgium, and France. There are few records from the Mediterranean. They occur off the eastern coasts of Canada and the United States and south to Florida, the West Indies, and the Bahamas. In the Southern Hemisphere, killer whales occur off the coasts of Australia and New Zealand. They have been seen off the coast of South Africa and near the Crozet Islands and the Kerguelen Islands and have been sighted in the Indian Ocean. They are in the South Atlantic and along the pack ice in Antarctic waters, where they occur as far south as the Ross Sea.

In the northeastern Atlantic, mating takes place in late autumn to midwinter. By the time they reach sexual maturity, males range from 5.2 to 6.2 meters in length; females, from 4.6 to 5.4 meters. Males are not monogamous; they may mate with more than one female during the mating season. Females (captive ones) give birth to their first calf after a long pregnancy of some 15 months when they are from 8 to 17 years of age. North Atlantic newborns may be as small as 183 centimeters in length; in the Southern Hemisphere, 227 centimeters. The mother whale supplies milk for about 12 months, but when the calf reaches about 4.3 meters in length, weaning begins; however, calves are dependent on their mothers for at least two years. Killer whale teeth erupt early, so it is not surprising that a captive calf started eating fish at the age of 11 weeks. Calves may be produced every 3 to 8 years, although the only killer whale in captivity to breed produced calves 19 months apart. The life span of killer whales in the wild is about 50 to 60 years for males and 80 to 90 years for females.

Killer whales consume large quantities of food. They are known to eat 4 percent of their body weight per day. Fishes, squids, and seabirds are primary elements of their diet, but more spectacularly they prey upon other large whales, seals, and dolphins and so have earned the name "killer." They are not fussy eaters; they are carnivorous and opportunistic and will eat whatever food is available in a region, which might differ from choice foods in another region. When the food supply moves, the whales may also migrate in response, moving with the food (see *What Do Whales Eat?*).

Most killer whale activities are oriented around the group; they are social animals normally found in pods of fewer than 40 individuals. Off the coast of Alaska, however, herds of up to 100 whales have been observed. A large herd is possibly a merging of several small ones, perhaps for the purpose of social activities or because it is the time of year when prey is available in quantity. Single whales are seen only infrequently. Pods consist of a mixture of adult males, adult females, and immature whales of both sexes.

The techniques used by killer whales to catch and kill their prey are notable because they are utilized in cooperative maneuvers by the pod. Indeed, these whales use strategy as wolves do in capturing prey. And what is more interesting is that they seem to teach these techniques to their young. They have been seen encircling schools of fish; then one or two whales will dash into the center and feed on the mass of food. In areas where prey occurs on ice floes, the whales might case the situation first by spy hopping; when they have located their prey, they surround the floe, inspect their prey, then back off and swim very fast toward the prey, dive, and cause huge waves to knock the seal or penguin off the floe into the waiting mouths of the whales. The whales may also hit the ice floe from beneath to knock off the pinniped or penguin. In other areas, killer whales have been seen to race toward the shore, partly stranding themselves so that they can eat seals at the surf line. In a pack, they have been observed pursuing gray whales, first in a straight line, then in a crescent-shaped formation, stationing themselves at 300-meter intervals, the better to catch their food. Killer whales have been photographed encircling, tearing at, and killing a group of nine female sperm whales. (See *What Do Whales Eat?*

and *Sperm Whales [Family Physeteridae].*) They are fearless and wily predators. Their only enemy is humans, who sometimes regard the presence of a killer whale as a competitor for food.

An unusual relationship developed between human and killer whale in the 1800s and 1900s in Australia. It was reported that, each July when humpbacks and right whales were passing through, killer whales would show up to alert shore-based whalers to the presence of whales close by. After killing the large whales, the whalers harvested the tongues of the dead beasts and gave them to the killer whales as a reward.

As awesome as killer whales are when hunting for food in the wild, in captivity they are easily trained and gentle. There have been some incidents where a trainer is shoved or pushed by a fluke or pulled underwater, but these are wild animals, so we cannot always predict when they might be frustrated, bored, feel aggressive, or are just overly playful. There is no documentation that the killer whale preys upon humans or has any interest in them as food.

Bottlenose Dolphin (*Tursiops truncatus*)   When most people think of the bottlenose dolphin they think of Flipper, the ambassador of the world of cetaceans. This species has been successfully kept in captivity since 1913 and has been the star of motion pictures and television shows. Although more is known about it that any other cetacean, we are still discovering many things about its natural history.

The bottlenose dolphin was brought to the attention of science in 1821 when George Montagu described an animal that was captured at Duncannon Pool on the south coast of England. Montagu was struck by the shortened or truncated nature of the teeth, which he compared to the molars of terrestrial mammals. He recognized that it was a dolphin and applied the generic name *Delphinus* to it, with the specific name *truncatus*, in allusion to its truncated teeth (Montagu 1821: 76). We now realize that this was an animal that was suffering from a dietary disease that rendered its teeth soft. By 1843 more of the diversity of dolphins was becoming known, so the generic name was changed to *Tursio* and in 1855 to *Tursiops*.

The bottlenose dolphin is an extremely variable animal, not only in its external appearance, but also in its skull and skeleton. These dolphins may be spotted or plain; the coloration may be light or dark gray; they may have long or short beaks and different numbers and sizes of teeth. The many variations have resulted in the application of 26 different species names to this genus. Recent studies have limited *Tursiops truncatus* to the offshore form found in the North Atlantic. The coastal form, from which most of the captive animals come, is probably a different species.

The bottlenose dolphin is distributed worldwide, from the equator to the cold temperate waters off the coast of New England, the United Kingdom, California, Chile, Argentina, Japan, and South Africa. It does not, however, swim into the Antarctic.

Three-quarter view of bottlenose dolphins in captivity. The usually large flippers that are characteristic of this species can be seen.

It is a moderately large dolphin, averaging 2.7 meters in length, with a maximum in the European population of 4.2 meters. A 2.7-meter adult bottlenose dolphin weighs about 200 kilograms. The bottlenose dolphin is subtly colored, and most people think of it as battleship gray. The belly is white, and faint stripes of lighter color occur around the eye, ear, and blowhole. The snout is long and robust, containing a total of about 80 to 100 teeth. The flippers are relatively large for a dolphin.

Tursiops becomes sexually mature at between 7 and 14 years, with an average age of 10. Their young are born in the spring, at a length of about 1.2 meters. The calves nurse for about 18 months but begin taking solid food at about 6 months.

Bottlenose dolphins are highly sociable animals, being found in groups of two or three up to herds of several hundred. Long-term studies on one coastal population have shown that the basic group tends to be a small number of adult females and their offspring. A number of other individuals irregularly join these groups, and the groups may join one another to form larger herds.

The bottlenose dolphin was first brought into captivity on 12 November 1913 at Cape Hatteras, North Carolina. There, an expedition from the New York Aquarium collected 10 dolphins from the bottlenose dolphin fishery. They were transported back to New York, where they lived in the New York Aquarium, which was operated by the New York Zoological Society on Coney Island. The bottlenose dolphin truly became a star with the advent of Marine Studios (later Marineland of Florida) in Saint Augustine, Florida. Trained dolphins were featured in public shows there in the 1950s, which brought the existence of the bottlenose to the attention of both the public and science. Behavioral experiments with the dolphins that led to the discovery of their sonar (echolocation) abilities were performed in 1952. Time passed and the public's enthusiastic reaction to captive dolphins increased, resulting in the establishment of a number of new oceanaria starring bottlenose dolphins. This wave of enthusiasm climaxed with the television show *Flipper* and the movie by the same name.

Considerable scientific work has been done on the bottlenose dolphin's ability to distinguish objects in its environment acoustically. This work has been successful because of the dolphin's adaptation to captivity. Recently, the intelligence of dolphins has generated much interest, again centered on the bottlenose. Fascinating work has been done, but the difficulty of comparing intelligence remains unresolved.

River Dolphins (Family Platanistidae)

Franciscana (*Pontoporia blainvillei*)   The scientific name for the franciscana is derived from the Greek *pontos*, for "open sea," and *poros*, for "passage" or "crossing," in the mistaken belief that this species lived in both freshwater and the ocean. The species name is derived from H. J. Ducrotay de Blainville, a well-regarded French naturalist. Its common name in Uruguay and Argentina, as in English-speaking parts of the world now, is "franciscana," and in Brazil, *toninha*. At one time the English common name used was "La Plata river dolphin."

Not well known, the franciscana is among the smallest of all cetaceans. Adult males range in length from 1.25 to 1.58 meters; adult females are larger, ranging from 1.34 to 1.74 meters. Indeed, the smallest brain of all cetaceans is that of an adult franciscana. One of the distinguishing features of the franciscana is its long slender beak and straight mouth line that curves gently upward near the ends. The beak is shorter in young animals, and the forehead bulges prominently upward, whereas in older animals the beak is longer and the forehead more rounded. The blowhole is crescent shaped with its open end toward the beak. The flippers are large and broad, like paddles; the flukes, pointed; and the dorsal fin, triangular with a rounded tip, stands moderately tall. Franciscanas are brown with shading that hints of a hood covering most of the dorsal surface. The ventral surface is a light gray. The teeth for this species are slender and pointed. In the upper jaw, they range in number from 53 to 58, and in the lower jaw, from 51 to 56.

Franciscanas inhabit the temperate, usually shallow coastal waters of eastern South America, from the Doce River, Brazil, to Peninsula Valdés, Argentina. Their diet consists of as many as 24 different species of bottom-dwelling fishes in addition to squid, octopus, and shrimp. Sharks are thought to prey on this species, although the killer whale, another predator, is also found in the franciscana's range. Remains of franciscanas have been found in the stomachs of seven-gilled sharks and hammerheads, but it is not known if the sharks caught the animals while they were free swimming or if they simply scavenged carcasses.

Generally solitary, franciscanas may sometimes be seen in groups of up to five animals. Females reach sexual maturity between the ages of two and four years. The smallest sexually mature female was 1.37 meters long and weighed 30 kilograms. The largest recorded female was 1.77 meters long. The oldest female examined by researchers was 13 years old; females may live only 15 years. Males, on the other hand, may live about 18 or 20 years. Males become sexually mature between two and three years of age. The smallest sexually mature male was 1.21 meters; the largest male was 1.58 meters long and weighed 32 kilograms.

Franciscanas give birth every other year, most occurring in November and December after a gestation period of 10.5 to 11.1 months. In Uruguayan waters, newborn are about .75 to .80 meter long and weigh about 7.3 to 8.5 kilograms. The calves drink mother's milk for about nine months, although at three months they begin to take some solid food. Females may be pregnant and lactating at the same time.

Sadly, the cause of excessive franciscana deaths can be attributed not to natural mortality or predation by its natural enemies but rather to the presence of human fishing activities. The main cause of death is entanglement in fishing nets. In the late 1960s and early 1970s, as many as 3,500 animals were caught incidentally in a gill-net fishery for sharks off Uruguay. Between 1971 and 1973, at least 536 dolphins were taken; between 1974 and 1978, a total of 1,394 were caught; between 1976 and 1985, 723 were found dead along a section of the Rio Grande do Sul coast. We know this must be just a tip of the iceberg, because statistics on incidental captures from other parts of the franciscana range are scarce.

This chapter has attempted to give the reader a brief glimpse into the evolution and diversity of whales and dolphins. We are just beginning to grasp the scope of diversification that fossil whales and dolphins went through. Treating the differences of the living whales and dolphins alone is complex enough. Because of the large number of living species (86), we felt we could not discuss them all in any kind of depth, so we chose species we feel are representative of their larger group.

.3.

WHALES AND HUMANS

DO WHALES ATTACK PEOPLE?

Large whales sometimes attacked whaleboats that provoked them, but we know of few events that could have been unprovoked attacks. The most famous attack was by a male sperm whale that rammed and sunk the Nantucket whaleship *Essex* on 20 November 1820. The historian Nathaniel Philbrick has recently completed a documentary account of the survivors' voyage. The accounts of the *Essex* were one of the inspirations for Herman Melville's story *Moby Dick*. In his 1878 book, *History of the American Whale Fishery*, Alexander Starbuck chronicles a number of accounts of merchant vessels that were sunk or severely damaged by whales (Starbuck 1878: 115). Most of these were probably accidental collisions. However, one recent account, by Dougal Robertson, documents a sailing ship that was sunk by a group of about 20 killer whales. The object of the attack in all cases seems to have been the vessel, because none of the persons involved was attacked.

An artist with one of Robert Scott's Antarctic expeditions was on relatively thin pack ice when a killer whale attempted to break through the ice under his feet. The artist managed to escape. One explanation for that whale's behavior is that possibly it saw the man's image through the ice and thought it was a seal. Killer whales have been documented tipping over ice floes or breaking them up to prey upon seals.

Certainly cases of captive whales interacting rather strongly with their trainers have occurred, but we have known for some time that captivity can disrupt the normal behavior patterns of animals. Whales and dolphins are large, fast animals. Any person interacting with them has to realize the dangers of inadvertent damage. Furthermore, it is illegal to feed or to swim with dolphins in the wild in some countries. One would be advised to observe cetaceans from a safe distance.

Narwhals, *Monodon monoceros*.

DO DOLPHINS PROTECT HUMANS FROM SHARKS?

One persistent tale implies that dolphins are enemies of sharks and will protect a human swimmer from a shark attack. There may be some correlation between the presence of dolphins and the simultaneous absence of sharks at times, but the correlation is tenuous. One of the coauthors (Mead) was swimming with dusky dolphins that were feeding off the coast of Argentina in 1973 when he was attacked by a 2-meter soupfin shark. The attack did not affect the dolphins' behavior, and the dolphins did not bother the offending shark. The author was chagrined to find that the dolphins did not carry him off to safety!

DO DOLPHINS SAVE PEOPLE FROM DROWNING?

Many stories about the interactions of dolphins and humans date back to the Greek legends about people riding on dolphins. However, we have no undisputed cases of dolphins saving swimmers from drowning. Usually there are no witnesses to support stories about this or any way to confirm these stories independently. Bottlenose dolphins frequent the surf zone, where swimmers may have difficulty. It is conceivable that a tired drowning swimmer would get thrown ashore by waves and think that the dolphins he had seen earlier had saved him. If dolphins simply have the tendency to push drowning swimmers, not necessarily toward shore, logically we could hear only from the lucky swimmers that were pushed toward the beach.

WHEN DID WHALING BEGIN?

Whaling began as soon as humans developed the means to kill whales. Prehistoric carvings and whaling artifacts demonstrate that whales were hunted long before recorded history. The earliest written records are in an account by King Alfred of Wessex (later called Alfred the Great of England) in the tenth century A.D. concerning Norse whaling. Alfred was referring to Ottar (or Othere, Ochther), who was a Norseman in the employ of Alfred and who voyaged around the North Cape (of Norway) to Perm (in what is now western Russia). In Ottar's description of his voyage, he gave an account of the whale and walrus hunting of the people known as the Biarmans (Permians), who lived on the shores of the White Sea. Ottar describes the Biarmans as taking whales that were "50 ells" in length. Depending upon the ell that Alfred used, the length of these whales was 30 to 70 meters. This is obviously a fabulous exaggeration, but nonetheless, it is indicative of very large whales.

Most people are unaware that much early whaling (by herding or driving whales ashore) took place without harpoons, lines, and floats. Some early whalers had discovered an extremely toxic plant poison called aconite, which was derived from wolfsbane or monkshood. The aboriginal inhabitants of Kodiak Island in Alaska used to shoot whales with arrows that were poisoned with aconite and then go back to shore and see if the whales would die and wash up. If the whales did not, the natives would wound other whales until they succeeded in harvesting an animal that drifted ashore. This was a potentially inefficient killing technique, in that many animals died without becoming available for food; thus, it risked the severe depletion of the whale population without productive results.

Drive fisheries have occurred many times. All that is essential is that the whale herds are capable of being driven in a predictable direction by the whalers. Usually the whales or dolphins are driven into a restricted bay, where they are killed. Pilot whales have been one of the classic species that is harvested this way. Pilot whales also mass strand, and many groups of them that were probably on their way to strand have been driven to another locality and killed.

Early Japanese whalers strung coarse nets along the migratory paths of whales. When the whales became entangled in the nets, people rowed out and harpooned them.

WHAT IS MODERN WHALING?

Modern whaling is executed from a boat equipped with an engine and a cannon that fires a harpoon with an explosive head. It was begun in 1865 by a Norwegian, Svend Foyn.

In early or aboriginal whaling, towing the whale with canoes could take many hours. Likewise, with early open-boat whaling, towing the whale back to the ship or station was time consuming. Even in modern whaling, towing the whales back to the whaling station with the catcher boat or tow boat can take up to a day and a half.

One development that led to severe overexploitation of the world's whale stocks was the factory ship. The earliest factory ships were just fishing ships that were installed with small steam-powered rendering works. The earliest factory ship is said to have been the bark *Laura*, which worked out of Sandefjord, Norway, in 1880. As whaling became more profitable, the industry used converted tankers that were installed with rendering works. Whales were flensed alongside the factory ship, which freed the whalers from the need to construct a land whaling station. Sheltered water was still necessary for flensing the whales, so whalers were restricted to the lee of islands and ice floes. In 1914 whalers ventured into the Antarctic, and in 1925 they developed the stern ramp, allowing them to bring the carcasses on board for flensing. This meant they were free to follow the populations of whales, which eventually led to serious overexploitation. In one year (the season of 1930–31), whalers killed 29,410 blue whales in the Antarctic, more than the current estimated world population of that species. At that time they were harvesting only oil. The meat was discarded.

WHAT TOOLS HAVE BEEN USED IN WHALING?

The first requirement in whaling is that whalers approach the whales closely enough to wound them. Primitive whalers set out in boats that were large and sturdy enough to protect them from the whale once it had been injured. These boats normally held from 6 to 12 men and were propelled by a combination of paddles, oars, and sails. Approaching the whale required stealth, often necessitating that the oars be muffled. When steam and diesel power became available and people designed special boats specifically to chase whales, then the stealth tactic was more or less abandoned. Modern designers still considered it, however, when they developed catcher boats with bronze propellers, which were thought to be better and quieter than steel propellers, even though the latter suffered less from contact with ice.

If the approach was successful, then whalers used various methods to wound the whale. Early aborigines used arrows and light spears dipped in poison, or they used slightly heavier harpoons that were tied to floats. The purpose of these floats was to tire the whale and impede its progress through the water. The whalers could then advance and kill the whale with lances. Open-boat whaling, which is often portrayed in movies, functioned in much the same way. The harpoon did not mortally injure the whale; it kept the whale attached to the boat. When the whale was worn out pulling the boat(s) along, the whalers drew close and killed it with lances.

It was not until the age of "modern whaling" with steam vessels and cannons that the harpoon actually killed the whale. Whalers had experimented with rocket-propelled harpoons in the middle of the nineteenth century, but that technique never proved commercially successful. The early harpoon cannons were of small bore, about 50 millimeters, which gradually increased to 90 millimeters, the modern standard. Early cannons were muzzle-loading, in that both the explosive charge that propelled the harpoon and the harpoon itself were loaded into the muzzle of the cannon. Because of its size, the harpoon always had to be loaded into the muzzle, but the explosive charge that propelled the harpoon eventually came to be loaded into a breech at the rear of the cannon. The harpoon normally carried an explosive head, which went off inside the whale and killed it.

Eskimo whalers flensing a bowhead whale. *Flensing* refers to the removal of the blubber only. The bowhead has relatively thick blubber, which provides insulation against the cold and gives the Eskimo the much-valued "muktuk" (blubber and skin) for eating.

WHAT IS FLENSING?

After whalers have killed the whale, they must transport it to the ship or whaling station, where it is taken apart. The operation of cutting up the whale is known by a word of Norwegian origin, *flensing*. It is usually done by one or two flensers, who wield the flensing knives. Winches are used to provide power for flensing. In some operations, notably factory ships, flensing is distinguished from lemming. The flensers are responsible for removing the blubber; the lemmers, for everything else.

WHAT ARE THE PRODUCTS OF WHALING?

In aboriginal, or artisanal, whaling, the carcass was consumed for meat and oil. In the seventeenth century, commercial whaling became established in Europe, engaging the efforts of the English, French, Dutch, Danes, Norwegians, Germans, and Portuguese. Their target species were the right whales (the black right whale and the bowhead, or the Greenland right whale, as it was known then), which were taken for lamp oil and, later, baleen. Before the advent of spring steel in the nineteenth century, baleen was used wherever there was call for a substance with elasticity (see *What Is Baleen?*). Fashions of that time were dependent on baleen for use as corset stays and stiffening for garments.

Modern whaling initially took baleen whales only for their oil. The oil of baleen whales came from rendering the blubber. This oil was used for human consumption, like margarine, and for industrial purposes such as making soap, paint, and high explosives. Whaling came almost to a standstill during the years of World War II. The demand for oil increased during those years, and vegetable oil made its appearance in the marketplace. After the war, modern whaling's primary purpose was to supply meat for human or pet consumption.

One whale product, not necessarily derived from a dead whale, is ambergris, which comes from the intestines of sperm whales and has been used as a perfume fixative (see *What Is Ambergris?*). Sometimes it is found floating in the ocean or washed ashore.

WHAT IS IVORY?

The use of the term *ivory* varies considerably. In the strictest definition, ivory is the dentine of the tooth of a mammal that has developed teeth large enough to encourage humans to hunt it. This, of course, includes the elephant, whose tusks form the ivory with which most of us are familiar. Other classical sources of that type of ivory have been the teeth of sperm whales and the tusks of walruses. Seals, bears, elk, and dolphins have also contributed their teeth to the ivory trade (see *Can We Identify Specific Varieties of Ivory?* and *What Whales Have Ivory?* for other sources). Some types of extremely dense bone have been used for the same purposes as tooth ivory (for decorations and tools) and have been called ivory.

Extending the definition further to include all materials of an ivory color and of sufficient hardness to be carved into ornaments, we include the nuts of certain plants. Tagua nuts have recently been popular. Material of this type is known as vegetable ivory.

Ivory is valuable, and there are, of course, many imitations of it. The first modern synthetic plastic, celluloid, was made from cotton fibers. It was used for imitation ivory in the late nineteenth century. The development of plastic polymers and modern casting techniques has led to their effective use as imitation ivory. We have seen several modern artifacts recently that we could not tell from ivory by visual examination.

CAN WE IDENTIFY THE SPECIFIC VARIETIES OF IVORY?

The larger an ivory piece is, the easier it is to identify the species from which it came. Carvings of sperm whale teeth in which the fine growth rings of alternating

The narwhal tusk has a unique external texture. The outside of the tusk consists of tooth cement, which is laid down in a spiral fashion. The spiral is always "left-handed," opposite the spiral of normal right-handed screws.

cement and dentine are evident can be recognized by the color differences between the layers (dentine is darker). Killer whale teeth are smaller but have the same characteristics as sperm whale teeth. Elephant ivory is composed of dentine that has a curious cross-hatched pattern in cross section, called the Schreger pattern. Different cross-hatched patterns allow fossil ivory from mammoths and mastodons to be distinguished from contemporary elephant ivory. Walrus tusks usually are composed of equal amounts of dentine and cement. The dentine in walrus tusks has the texture of coarsely crystalline sugar.

Some pieces of ivory retain the overall external appearance of the tooth or traces of its surface. For example, boar tusks are markedly curved, and the surface of hip-

TABLE 6. KINDS OF IVORY MATERIALS, WITH TYPICAL DIMENSIONS

Tusks
    Narwhal (4–6 cm × 1+ m)
    Walrus (4–5 cm × 30+ cm)
    Elephant (10–20 cm × 1.5+ m)
    Hippopotamus (4–6 cm × 30 cm, strongly curved and grooved)
    Boar (1–2 cm × 10–12 cm, strongly curved)
Teeth
    Sperm whale (6–8 cm × 10–20 cm)
    Killer whale (2–4 cm × 8–12 cm)
    Walrus (1–1.5 cm × 2–3 cm)
    Elk (1–1.5 cm × 2–2.5 cm)
Other
    Antler
    Bone
    Tagua nut (2–4 cm)

popotamus tusks is coarsely grooved. Narwhal ivory is extremely easy to recognize because of the pronounced spiral in the outer cement layer.

Bone that has been used as ivory can sometimes be distinguished by type. The type that can be identified by the pattern formed by small vascular Haversian canals is extremely dense (compact or cortical) bone. In less dense (medullary or cancellous) bone, the calcium struts of bony supporting trebeculae can be seen.

If the piece of ivory is small and uniform in coloration, it could be difficult to identify visually. Putting a small sample in a gas flame can be diagnostic, particularly of fake ivory. Bone or teeth give a distinctly organic odor, similar to burning hair; plastic produces a strong synthetic odor. Usually burning is not an option, though, because of the amount of material it consumes, and owners may not want the specimen damaged.

WHAT WHALES HAVE IVORY?

Sperm whales were the principal source of ivory for whalers' scrimshaw. Sperm whales have 17 to 30 pairs of teeth in the lower jaw and none erupted in the upper jaw. These teeth can be up to 30 centimeters long and 10 centimeters thick. Internally the teeth consist of dentine; externally, of cement. Dentine is darker than cement, and it is impossible to cut or carve large items from sperm whale teeth without revealing that difference in color. Thus sperm whale teeth did not compete with elephant tusks in the piano key and pool ball markets.

Whalers were always looking for other whale parts that could be used for carving, particularly elements that were longer than a sperm whale's tooth. They sometimes carved the posterior portion of the jaw of sperm whales, which was known as the pan bone, and carved walking canes out of the middle bone of the rostrum (premaxilla) of baleen whales.

Narwhals have uniquely spiraled tusks that are nearly 3 meters long and 8 centimeters in diameter. These have been long used as ivory ornaments and trophies, including a Danish throne that was constructed of narwhal tusks.

WHAT IS SPERMACETI?

Spermaceti is a sperm whale–derived substance that has been used for making waxes, candles, and cosmetics. In fact the definition of one candlepower (the international unit of illumination) is the amount of light produced by a candle made of pure spermaceti burning at the rate of 120 grains (7.776 grams) of wax per hour. Spermaceti was one of the principal economic factors of the sperm whale fishery during the middle of the nineteenth century, when the British and Americans were active in it (see *Why Did Whaling Decline?*).

Spermaceti was mistakenly thought in the sixteenth and seventeenth centuries to be the sperm (semen) of whales, hence its name. However, spermaceti is actually a mixture of wax and oil that is contained in the spermaceti organ, or "case," of sperm whales. *Case* is a term used by whalers to refer to the enormous connective-tissue structure that forms the top (dorsal) portion of the head of sperm whales. This structure can be up to 6 meters long. It consists of a series of cavities that contain up to 2 metric tons of spermaceti. Scientists now think that the spermaceti organ has an acoustic function (see *Do Whales Stun Prey with Sound?*).

Spermaceti might enhance diving through its temperature changes in the organ. During the early stages of a dive, the oil would cool down and increase in density, hence decreasing the whale's buoyancy. When the whale surfaces, the oil would become warmer, expanding in density and increasing the animal's buoyancy. The spermaceti organ also possibly facilitates the emptying of the lungs and absorbs excess nitrogen during long dives.

The spermaceti organ is unique to the sperm whale family, including dwarf and pygmy sperm whales. It probably derived from a small structure (the right posterior dorsal bursa) in the forehead of other toothed whales. The remainder of the toothed whale species do not have a spermaceti organ, but they do have a "melon," which contains a substance that is remarkably similar to spermaceti. *Melon* is the term used by early whalers who took pilot whales, in which this structure is particularly pronounced and looks like a section of cantaloupe. It is, however, in the forehead of all toothed whales, including sperm whales, where it is known as the junk. The melon or junk is composed of fat and connective tissue.

WHY DID WHALING DECLINE?

Principally whaling declined because of overexploitation of whale stocks and their subsequent protection and because it ceased to be economically important. With the decline of the profitable right whale stocks in the late eighteenth century and the shifting of the maritime powers in the early nineteenth century, the United States and Britain dominated the whale fishery with voyages that lasted about four years and roamed the world's oceans looking for sperm whales. The sperm whale fishery took whales for oil and spermaceti, a substance that is a combination of oil and wax, found in the head of the whale. The oil was burned in oil lamps, and spermaceti made the finest smokeless candles. The U.S. whaling fleet was virtually eliminated in the Civil War. That, coupled with the discovery of petroleum in

Pennsylvania, eliminated the need for sperm oil and spermaceti and put an end to Yankee sperm whaling.

The next phase in whaling history was the advent "modern whaling." This period began with a local Norwegian whaler's invention of a whaling cannon. Devised by Svend Foyn in 1865, this weapon was a muzzle-loading "3-inch cannon" (the name derives from the bore diameter) that fired an explosive harpoon and was mounted in the bow of a small, newly designed, steam-powered catcher boat. That made possible the taking of large fin and blue whales, which had heretofore been safe from exploitation. Modern whaling led to such a depletion of stocks that it was forced into decline.

WHEN DID WHALING DECLINE?

Modern whaling began to decline in the 1960s with increased exploitation and the subsequent failure of stock after stock of whales. At the present time, it is practiced by very few countries, and the take is greatly reduced.

WHAT ABOUT DOLPHIN FISHERIES?

Dolphins have been captured and consumed wherever humans have had the technology and occasion to do so. Usually their take has been incidental to other fisheries' catches, but sometimes a dolphin (or porpoise) fishery has existed on its own. In northern Europe and North America, particularly in Norway, Iceland, and Newfoundland, dolphins and porpoises were commonly taken. Several small fisheries for pilot whales existed in those countries. A drive fishery for the pilot whale (also a member of the oceanic dolphin family) was practiced on Cape Cod until the 1930s. The only takes of dolphins and pilot whales in the United States were primarily for the oil that could be found in the heads (melon oil). This oil was used for lubrication of watches and sewing machines. Fisheries for the bottlenose dolphin existed at Cape May and Cape Hatteras. The Cape May fishery also processed the dolphin hides for leather. The last recorded captures were in 1928 at Cape Hatteras.

ARE WHALES AND DOLPHINS KILLED ACCIDENTALLY?

A major problem with fisheries of the world is their "by-catch," or the animals other than the target species that the fisheries take. By-catch can include whales, dolphins, porpoises, seabirds, and many species of fishes, which may simply be discarded at sea. The most infamous example of this was the routine by-catch of dol-

Left: Developing countries still use all the protein they can get, including these young spinner dolphins photographed in a fish market in Sri Lanka. *Right:* The utilization of dolphins and porpoises as meat is not restricted to developing countries. The Dall's porpoises shown in this photograph were taken in a harpoon fishery in northern Japan.

Dolphins and porpoises are also taken incidentally by fisheries whose target species are fishes. This photograph shows the catch of a gill net that was set for *totoava*, a fish that is esteemed in the markets of the Gulf of California. The catch consisted of fishes, a large shark, and a critically endangered vaquita, or Gulf of California harbor porpoise.

phins (mainly spotted and spinner dolphins) by the American tuna fleet. The tuna by-catch has been decreased in some countries but is still a problem with boats of nonregulated countries.

Any fishing method that catches animals without a human decision as to which individuals are taken (methods such as gill-nets, trawls, seines, long-lines, etc.) is a potential problem. Only techniques that utilize individual fish catches, such as hook-and-line tuna fishing (of old-style tuna clippers) or swordfish spearing, do not have a by-catch problem. As the equipment gets bigger and bigger, so does the problem. Fortunately international agreements started curtailing the pelagic drift gill-net fishery because of its by-catch. That fishery used nets up to 50 kilometers in length and 12 meters in depth.

HOW ARE CETACEANS PROTECTED?

More than 2,033,244 whales in the Antarctic—an enormous number—were exterminated in the years 1904–93 by the whaling activities of the various nations of the world. Of these, 359,885 were blue whales, 394,672 were sperm whales, and 723,340 were fin whales. The extinction of the Atlantic gray whale in the early 1700s has been attributed to human activities. The northern right whale of the western North Atlantic is barely holding on, with only about 300 individuals surviving. Whaling still continues. We have had one success in rejuvenating cetaceans' numbers—the rebound of the California gray whale. So the question remains, How do we best protect and conserve these magnificent creatures for the generations to come? Increasingly, people find themselves in conflict with other species for space and resources; therefore, a commitment must be made to balance our needs with those of the natural world. The only hope for the preservation of whales is international cooperation.

If we look back in history to the early part of the twentieth century, when attention was finally focused on the plight of whales, we see the beginning of a series of conventions, international agreements, treaties, and various types of national and international legislation enacted to restrict whaling and protect those species in imminent danger of extinction.

The First National Legislation   Modern whaling had begun only four decades before the early part of the twentieth century, when we first became concerned about the possible extinction of many species. Because modern whaling started in Norway, that country had the first national legislation attempting to govern whaling. In 1880 Norway enacted the Whale Protection Act, at least in part to control Svend Foyn, the originator of modern whaling. Later, the Norwegian Whaling Act

A dolphin in a harbor along the Texas coast. One cannot help but think that the life of whales and dolphins in such intensively utilized habitats as this is a step in the wrong direction.

of 1929 established a duty of 20 ore per barrel of whale oil, to be applied to the expenses connected with the Whale Protection Act.

International Convention for the Regulation of Whaling   National conservation efforts begun in the United States, Great Britain, and Norway resulted in the League of Nations' convening a conference in Berlin in June of 1930. This resulted in the document Draft Convention for the Regulation of Whaling, which became official in 1931 and was put into force in 1935. However, that effort did not restrict whaling enough, so concerned nations held a series of conferences in London in 1937 and 1938, resulting in the International Agreement for the Regulation of Whaling. As with any international agreement for the regulation of a valuable resource, there were abundant loopholes. World War II afforded the only real protection of the whales.

After the end of World War II, in November 1946 the International Whaling Conference was held in Washington, D.C., which drew up the International Convention for the Regulation of Whaling and laid the framework for the International Whaling Commission. The conference also drew up the Protocol for the Regulation of Whaling, which was to apply to the 1947–48 season. The International Whaling Commission was actually formed at a meeting in London in 1949. Actual quotas on the whales allowed to be taken were set and published in the Schedule of the International Convention for the Regulation of Whaling.

U.S. Marine Mammal Protection Act of 1972   The Marine Mammal Protection Act of 1972 (MMPA) was most recently reauthorized in 1994. This act is a legislative landmark, because it strives to protect not only all species of marine mammals but their environment as well. If an action can be linked to harmful effects on marine mammals, such as a fishery that is competing with marine mammals for fish, the fishery becomes subject to enforcement action.

In passing the MMPA in 1972, Congress found that

> certain species and population stocks of marine mammals are, or may be, in danger of extinction or depletion as a result of man's activities; such species and population stocks should not be permitted to diminish beyond the point at which they cease to be a significant functioning element in the ecosystem of which they are a part, and, consistent with this major objective, they should not be permitted to diminish below their optimum sustainable population level; measures should be taken immediately to replenish any species or population stock which has diminished below its optimum sustainable level; there is inadequate knowledge of the ecology and population dynamics of such marine mammals and of the factors which bear upon their ability to reproduce themselves successfully; and marine mammals have proven themselves to be resources of great international significance, aesthetic and recreational as well as economic.

The MMPA established a moratorium, with certain exceptions, on the taking of marine mammals in U.S. waters and by U.S. citizens on the high seas and on the importing of marine mammals and marine mammal products into the United States.

Under the MMPA, the secretary of commerce is responsible for the conservation and management of cetaceans and pinnipeds (other than walruses). The secretary of the interior is responsible for walruses, sea and marine otters, polar bears, manatees, and dugongs. The secretary of commerce also delegated MMPA authority to the National Marine Fisheries Service. Part of the responsibility that the service has under the MMPA involves monitoring populations of marine mammals to make sure they stay at optimum levels. If a population falls below its optimum level, it is designated as "depleted," and a conservation plan is developed to guide research and management actions to restore the population to healthy levels.

Also under the mantle of the MMPA, the Marine Mammal Commission compiled a set of volumes consisting of 4,564 pages to list the agreements dealing with marine mammals. Among these agreements are quotas established by whaling countries (the United States, Japan, Greenland, Iceland, Norway, and Russia) that are intended to limit the take of whales.

U.S. Endangered Species Act The majority of the large whales are designated by the United States as endangered species and, as such, are covered by the U.S. Endangered Species Act (ESA). The ESA not only protects them from direct harm but also generates research funds so that scientists can learn more about them.

The Endangered Species Act of 1973 reads:

> The purposes of this Act are to provide a means whereby the ecosystems upon which endangered species and threatened species depend may be conserved, to provide a program for the conservation of such endangered species and threatened species, and to take such steps as may be appropriate to achieve the purposes of the treaties and conventions set forth in subsection (a) of this section. These are:
>
> (A) migratory bird treaties with Canada and Mexico;
> (B) the Migratory and Endangered Bird Treaty with Japan;
> (C) the Convention on Nature Protection and Wildlife Preservation in the Western Hemisphere;
> (D) the International Convention for the Northwest Atlantic Fisheries;
> (E) the International Convention for the High Seas Fisheries of the North Pacific Ocean;
> (F) the Convention on the International Trade in Endangered Species of Wild Fauna and Flora; and
> (G) other international agreements.

All species of cetaceans that were subject to commercial whaling in 1973 were originally listed as endangered under this act. This left out the minke and Bryde's whales. The California population of the gray whale has recovered sufficiently, and its listing has been changed from endangered to depleted. The Yangtze river dolphin (baiji), the Indus river dolphin, and the vaquita (Gulf of California harbor porpoise) continue to be listed as endangered.

Convention on the International Trade in Endangered Species of Wild Fauna and Flora In the last half of the twentieth century, international concern increased regarding the failure of attempts on a national basis to regulate trade in endangered species. In 1973 a meeting was called to frame the Convention on the International Trade in Endangered Species of Wild Fauna and Flora (CITES), which was signed on 3 March of that year. According to that convention, any trade in endangered species may operate only when *both* the country of origin *and* the country of destination formally issue permits to allow it. It entered into force after the tenth ratification on 1 July 1975 and has significantly decreased trade in items such as elephant ivory.

The list of species covered under CITES is given in three appendixes.

Appendix I shall include all species threatened with extinction which are or may be affected by trade. Trade in specimens of these species must be subject to particularly strict regulation in order not to endanger further their survival and must only be authorized in exceptional circumstances. The export of any specimen of a species included in Appendix I shall require the prior grant and presentation of an export permit. The import of any specimen of a species included in Appendix I shall require the prior grant and presentation of an import permit and either an export permit or a re-export certificate.

Appendix II shall include: (a) all species which although not necessarily now threatened with extinction may become so unless trade in specimens of such species is subject to strict regulation in order to avoid utilization incompatible with their survival; and (b) other species which must be subject to regulation in order that trade in specimens of certain species referred to in sub-paragraph (a) of this paragraph may be brought under effective control. The export of any specimen of a species included in Appendix II shall require the prior grant and presentation of an export permit. The import of any specimen of a species included in Appendix II shall require the prior presentation of either an export permit or a re-export certificate.

Appendix III shall include all species which any Party identifies as being subject to regulation within its jurisdiction for the purpose of preventing or restricting exploitation, and as needing the co-operation of other Parties in the control of trade. The export of any specimen of a species included in Appendix III from any State which has included that species in Appendix III shall require the prior grant and presentation of an export permit. The import of any specimen of a species included in Appendix III shall require, except in circumstances to which the paragraph below of this Article applies, the prior presentation of a certificate of origin and, where the import is from a State which has included that species in Appendix III, an export permit.

In the case of re-export, a certificate granted by the Management Authority of the State of re-export that the specimen was processed in that State or is being re-exported shall be accepted by the State of import as evidence that the provisions of the present Convention have been complied with in respect of the specimen concerned.

Considerable protection has also been given whales by the establishment of sanctuaries, or protected areas. One area where protection is common is the ocean around Antarctica. Much of Antarctic wildlife is covered by international agreements, such as the Convention on the Conservation of Antarctic Marine Living Resources (CCAMLR). Whales are also protected by a number of state laws, which are usually more specific than international agreements. For example, the state of Washington has a law prohibiting the live capture of killer whales in Puget Sound. Whales have become an icon of conservationists.

WHAT IS THE IWC?

In the late 1920s, the take of whales in the southern oceans became so great that scientists were concerned about the ability of the populations to withstand it. The first measure aimed at restriction of whaling activities was the Convention for the Regulation of Whaling, in Geneva in 1931. That convention was drawn up by the League of Nations and dealt only with baleen whales. The league was in the process of failing, and therefore the convention was of no practical value. This was followed by the International Agreement for the Regulation of Whaling, in London in 1937. Then World War II broke out, and whaling ceased. The London agreement did not have time to work.

Shortly after the end of World War II, a meeting of delegates took place in Washington, D.C., to draw up the International Whaling Convention. The convention was signed on 2 December 1946 and went into effect in 1948, establishing the International Whaling Commission (IWC), which was, at the time of its formation, composed of representatives of whaling nations. The commission designated a scientific advisory committee for guidance in regulating the take of whales, but conservationists faulted the commission for being composed of persons who had an interest in whaling. Despite this criticism, the IWC was certainly an improvement over no regulation at all. It was the first major international body to conserve populations of a natural resource. In the 1970s, the IWC opened up its membership to all countries, and in the 1980s it instituted a global moratorium on whaling among its member countries.

WHAT IS WHALE WATCHING?

Watching whales has, in the last few decades, grown into an important activity, both in terms of people's interest and in terms of the amount of money that it attracts. Whale watching began on the Pacific coast of the United States in the 1960s, when people were alerted to the annual migration of the gray whale. As time went on, and with more media coverage, more people began to discover that local populations of different species of whales were accessible from almost every port in the world. Whale watching is also a "cash venture" for traditional whaling

One of the positive things that has come out of the last 30 years is the development of whale watching. Overzealous whale watching has some possible deleterious effects, but the growing appreciation of whales as a part of the natural world more than makes up for them.

countries, such as Japan and Norway. The operators of whale-watching boats have a legitimate argument that the whales are worth more alive than dead. Whale-watching has made important scientific contributions. For example, methods for recognizing individual whales grew out of whale-watching endeavors.

HOW HAVE WHALES FIGURED IN FOLKTALES AND THE ARTS?

The northern European unicorn legend, involving a mythical creature with the body of a horse with one long horn on its forehead, had the narwhal as one of its bases. Tusks of the narwhal had long been used as trade items from the Far North. For the ordinary medieval European, who was not familiar with the narwhal, the tusks represented fantastic animals, and the traders encouraged that belief. In the Europeans' struggle to understand the origin of the narwhal tusk, they substituted it for the horns of a Near Eastern antelope, the oryx, in their thinking. Then they

The star of the story of *Moby Dick* was an albino sperm whale. Much has been written about the possible existence of albino sperm whales, and this photograph shows an "albinistic" (nearly all white), if not albino, sperm whale calf.

A group of three narwhals with their tusks emerging. The narwhal was one of the animals that served as a basis for the unicorn legend. Medieval Scandinavian traders fostered the stories of this mythical animal in order to increase the value of the narwhal tusks that they had to trade.

One of the favorite subjects of scrimshaw artists was the vessels they were on, here the *Charles W. Morgan,* carved on a sperm whale's tooth.

imagined the body as that of a more familiar animal, the horse. But the narwhal tusk was the only tangible evidence for the medieval illustrators, who accurately depicted the proportions and the left-hand spiral of the tusk. One suspects, however, that the Danes, who carved a royal throne out of narwhal tusks, must have understood the origin of them.

Whalers seem to have made most of their contributions to songs and ballads in the early days when they were hunting in the North Atlantic for the Greenland right whale (now known as the bowhead whale) or cruising around the world after sperm whales. The crews are known to have sung shanties during their long hours of toiling at flensing the whales, but they seemed content to repeat the few old traditional themes. Despite the number of whalers that served in Yankee sperm whaling, they seem to have come up with relatively few songs. The following are some of the whalers' contributions to our musical heritage:

GREENLAND WHALE FISHERY
Our captain stood on the fo'cas'le head
With his spyglass in his hand,
"There's a whale, there's a whale, there's a whale," cried he,
"And it blows on every span, brave boys,
And it blows on every span."

BLOOD RED ROSES

As I was going round Cape Horn,
Go down, you blood red roses, go down.
I wished to the Lord I'd never been born,
Go down, you blood red roses, go down.

    O you pinks and posies,
    Go down, you blood red roses, go down.

Around the Cape in heavy gales,
And it's all for the sake of them sperm-whales,
I wished to the Lord I'd never been born,
Go down, you blood red roses, go down.

    O you pinks and posies,
    Go down, you blood red roses, go down.

GREENLAND'S ICY SHORES

The weeks passed by and the days rolled on,
    And ne'er a whale did we spy,
'Til the lookout, he cried, "Whar away?" and
    "Thar she blows!"
    Far ahead and a little on our lee, brave boys,
Far ahead and a little on our lee.

BEHRING SEA SONG

Full many a sailor points with pride,
    To cruises o'er the ocean wide,
But he is naught compared to me,
    For I have sailed the Behring Sea.

We breakfast, dine, and sup on fat,
    Eat walrus blubber and all that;
Bull seals and whales are our delight,
    And polar bears we love to fight.

THE BO'S'N

"We sail'd and sail'd and one fair moon,
    A great whale we espied;
So we took a rope and a long harpoon,
    And stuck him in the starboard side.
Then away and away went the great big whale,
    And away and away went we;

Tied fast to his tail to the North we did sail,
    And that's the truth," said he;
"Tied fast to his tail to the North we did sail,
    And that's the truth," said he.

BLOW, YE WINDS
IN THE MORNING

'Tis advertised in Boston, New York and
    Buffalo,
Five hundred brave Americans, a-whaling for
    to go.
Singing:
Blow ye winds in the morning, blow ye winds,
    high-o!
Clear away your running gear, Blow ye winds,
    high-o!

THE WHALE

Now the boats were launched and the men
    aboard,
    With the whalefish full in view,
Resolved were the whole boats' crews
    To steer where the whalefish blew,
    brave boys,
To steer where the whalefish blew.

Whaling gave rise to one of the great classics of American literature, *Moby Dick*, written by Herman Melville and first published in 1851. Melville was a prolific author, and his stories take place principally in maritime settings. None is as famous as his allegorical novel *Moby Dick*, based on his time serving aboard a whaler in 1841. The story tells of Ahab, captain of the whaling vessel *Pequod*, who seeks vengeance on a white sperm whale whom he blames for intentionally causing the loss of his leg. Ahab lost his leg after it had been damaged by the whale on a previous voyage, and he is fanatically determined to wreak revenge on his nemesis and destroy him. *Moby Dick*, with its abundant symbols, allusions, and eloquent language, has been praised as describing man's struggle against the elements (or natural forces) as well as man's struggle with his conscious and unconscious self. There are many interpretations of Melville's story, which we will not dwell upon in this text. Nevertheless, we spotlight his chapter "Cetology" as a superb biological treatment of what was known in 1851 about whales.

Moby Dick was based on a number of historical events, the two most influential being Melville's time spent working on a whaler in 1841 and the sinking of the *Essex* by a sperm whale on 20 November 1820, as told by the mate Owen Chase. The *Essex* incident demonstrated that whales were capable of sinking whaling vessels. The actual whale on which Melville based Moby Dick seems to have been a composite based principally on stories about "Mocha Dick," a white sperm whale in the Pacific in the early part of the nineteenth century. A true albino sperm whale, with pink eyes and no trace of pigmentation in its skin, has yet to be seen by scientists. There are enough records of "albinistic" (pinto) sperm whales to make it clear that a whale more white than black could exist.

Other books on whaling can best be described as documentaries or technical treatises, some of which rate as literature. One frequently overlooked work is *The Cruise of the Cachalot*, by Frank Bullen. Based on Bullen's personal experiences, this book is a highly readable, accurate, first-hand account of whaling. Another interesting documentary is Erle Stanley Gardner's book *Hunting the Desert Whale*, a story of an early whale-watching trip to Baja California. One complete work of fiction based on modern whaling is Laurens Van Der Post's *The Hunter and the Whale*.

Movie scripts about whales or whaling have fared poorly compared with more popular themes featuring cowboys, truckers, and soldiers. Undoubtedly the one great whaling movie is *Moby Dick*, directed by John Huston and starring Gregory Peck. Possibly the costs of producing a movie with a whaling setting outweigh the cash receipts at the box office. The animated portion of the cinematic world has fared just as poorly. Nevertheless, there are the unforgettable cetacean characters, such as "Monstro," the whale in Walt Disney's *Pinocchio*, who was a hybrid of a blue and a sperm whale (it was blue and had ventral grooves, but it also had a mouthful of teeth), and another fabulous hybrid, "Willie," of a forgotten Disney special entitled *The Whale Who Wanted to Sing at the Met*. Live trained whales have contributed such characters in nonanimated features as the killer whale in *Free Willie* and the fin whale in *A Whale for the Killing*.

Scrimshaw is a term that is used for both the act of engraving images on a piece of shell, a whale's tooth, a walrus tusk, or a similar object and the artwork that results from it. It was practiced principally by whalers who had ready access to the raw materials. The definition of scrimshaw is broad, and so the term has been applied to any decorative engraving, even the intricate engravings that were sometimes produced by the workshop of Stradivarius on the bodies of violins. The use of the term *scrimshaw* has sometimes been expanded to cover any artwork that was done by whalers, whether it involved engraving or not. One book defines scrimshaw as "whaleman's or sailor's art of decorating whalebone and ivory with scratches to form pictures or designs, also attributed to primitives of various tribes" (Meyer 1976: 262). Scrimshaw bloomed in the early part of the nineteenth century, when Yankee whaling for sperm whales was at its peak. For this reason most definitions strongly imply that scrimshaw was done specifically by *American* sailors or whalemen and on *sperm whale teeth.*

In fact, the most typical form of scrimshaw did use a tooth of a sperm whale, which was sanded on at least one side to form a smooth surface and was then engraved. The whaler then worked an image into the tooth with an engraving tool (a knife, a needle, or a pick). The image consisted of a series of fine lines that were subsequently filled with lampblack to create contrast. The lines of some pieces of scrimshaw were filled with different colors, and some pieces were painted. The feature that they all had in common was that their images were engraved.

Scrimshaw has become a collector's item, and several museums in the northeastern United States specialize in collecting and exhibiting it, such as the Kendall Whaling Museum in Sharon and the Old Dartmouth Historical Society in New Bedford, Massachusetts. Our thirty-fifth president, John F. Kennedy, was an avid collector of scrimshaw.

IS THE STORY OF JONAH AND THE WHALE TRUE?

We will never know if Jonah was actually swallowed by a whale. If a person is asking, "Can anyone survive 72 hours in the stomach of a whale?" the answer is no. Jonah is a biblical character in the Old Testament (the Book of Jonah). He was thrown into the sea by sailors after he had told them he was responsible for a storm that had beset them because he had angered God. "Now the Lord had prepared a great fish to swallow up Jonah. And Jonah was in the belly of the fish three days and three nights" (Jon. 1:17, King James Version).

"A great fish" is commonly interpreted to mean a whale. In Hebrew it is shown as a *dag gadol. Dag* has a broad meaning encompassing anything that lives in the sea. The common argument made against Jonah's actually having been swallowed by a whale involves the misconception that a whale's gullet (esophagus) is not wide enough to permit a man to pass through it. This misconception is based on early naturalists' observations that the esophagus of a big baleen whale is about 10 centimeters in diameter, but they did not realize that the esophagus is extremely elastic and clearly capable of stretching enough to admit a human. The gullet of the biggest toothed whale, the sperm whale, is easily wide enough. There are several stories of whalers who were swallowed by sperm whales and were found when the whale was flensed. In most of the stories the whaler was said to have died, but some stories tell of the recovery of a living sailor who had spent several hours inside the whale. These stories try one's faith. There is no source of air in a sperm whale's stomach, and it is highly improbable that a person could survive more than a couple of minutes without suffocating.

WHAT IS LEFT TO LEARN ABOUT CETACEANS?

Much remains to be learned about cetaceans' ancestors, their current geographic distribution, and even their anatomy. The fossil record for whales has barely been touched. New discoveries are being made every year, and we are slowly beginning to appreciate the diversity of fossil forms. That relates to the question of the systematic relationships (who is related to whom?) of the fossil and recent species. Discoveries and new techniques, such as DNA sequencing, permit greater insight into the origin of the living species. We know that unnamed species are presently living in the oceans. *Mesoplodon pacificus* is known only from two skulls; we do not even know what the whole animal looks like. There is another species of beaked whale in the Pacific Ocean that has been sighted and appears to be a new form. The most recent new species of whale (*Mesoplodon bahamondi*) was named in 1995.

Much remains to be learned about the geographic distribution of whales, particularly what factors enter into the distribution of a population. Those species of whales that mass strand have strayed from their normal habitat, and we do not know why or even what their normal habitat is. We know the distribution of most whales for only a part of the year. We know, for example, that fin whales come into Newfoundland coastal waters in the spring and leave in the fall, but we do not know where they live in the winter. There will always be questions about the abundance of whales as time goes on and about the effect of humans and other ecologic factors on their population changes.

Much also remains to be learned about the anatomy of cetaceans, both gross and microscopic. Because we have so little opportunity to observe the habits of most whales, we attempt to learn a little bit about their habits by analyzing their functional anatomy. For example, work done on the anatomy of the muscles that function in swimming could reveal differences in swimming speed among the major groups of cetaceans. We need more work on the physiology of whales, particularly with regard to the senses of taste and smell. We also need to do more work on disease and parasitism, particularly on the parasites themselves, to understand the origin of the common parasites and their effects on whales and possibly other animals. Our work on the behavior of cetaceans increasingly needs to be done in such a manner that comparisons can be made among species and populations.

WHAT IS THE PROGNOSIS FOR WHALES AND DOLPHINS?

A number of cetacean species are coming back, but the North Atlantic and North Pacific right whales are not. It seems that the slow-swimming right whales are more subject to adverse human interaction, such as collisions with vessels and entanglements in fishing gear.

Dolphins and porpoises also have their problems with humans. Up until the 1920s they were actively harvested for their hides and oil. When this became unprofitable, they had a brief breather until large-scale fishing efforts commenced to catch them by accident. Now it seems that when one fishery stops killing cetaceans, another one springs up.

There is hope for the species of whales and dolphins whose geographic ranges are large enough to permit them to survive. The species that are in dire trouble now are those with a limited geographic range. The vaquita, a porpoise that lives in the northern part of the Gulf of California, is barely hanging on, because it is subject to gill-net fisheries and ecological disturbances from the badly polluted Colorado River. Similar situations occur with the Yangtze river dolphin (baiji), the finless porpoise, the Ganges and Indus river dolphins, and the Amazon river dolphin (boutu).

Of course, humans continue to damage the environment with pollution. In some enclosed bodies of water that are adjacent to heavy industrial centers, such as the Baltic Sea, the effects of pollution on marine mammals are seen in external sores and problems with reproduction. Closer to home, the beluga population in the Saint Lawrence River is suffering from the effects of pollution. Even where there is no demonstrable harm, we cannot help but think that pollution increases disease and decreases longevity.

Still, there is hope. The California gray whale, thought to be extinct in the early 1900s, has come back. Since large-scale commercial whaling on it ceased, its population has increased to about 26,000 at the beginning of the twenty-first century.

HOW CAN SOMEONE BECOME A WHALE EXPERT?

A career working with whales can be very rewarding, but it involves a great amount of effort. It must begin with a broad-based education in biology and then pursuit of a graduate degree in a field that can be applied to cetaceans. Appendix 2, Careers in Marine Mammal Science, should help those seeking to become a cetacean biologist or other type of cetacean professional. Of course, it is not necessary to work with whales to appreciate them. The sources of information outlined in Appendix 3 provide an avenue for anyone wishing to advance his or her knowledge about whales.

APPENDIX 1.

GENERAL CLASSIFICATION OF MAMMALS AND SPECIFIC CLASSIFICATION OF WHALES AND DOLPHINS

GENERAL CLASSIFICATION OF MAMMALS

Humans are classified as belonging to the species *Homo sapiens* of the family Hominidae, order Primates, subclass Placentalia, class Mammalia, and phylum Chordata. Whales belong to the same class, Mammalia, but to a separate order, the Cetacea. The following is a partial classification of Mammalia designed to show the relationships of the whales to other mammals.

CLASS MAMMALIA (mammals—animals with fur)
Subclass Monotremata (monotremes—platypuses and kin)
Subclass Marsupalia (marsupials—opposums and kangaroos)
Subclass Placentalia (placental mammals)
  Order Xenarthra (sloths, armadillos)
  Order Insectivora (shrews, moles)
  Order Scandentia (tree shrews)
  Order Dermoptera (flying lemurs)
  Order Chiroptera (bats, fruit bats)
  Order Primates (monkeys, apes, and man)
  Order Carnivora (dogs, cats, seals, bears, hyenas)
  Order Cetacea (whales and dolphins)
    Suborder Mysticeti (baleen whales)
      Family Balaenopteridae (fin, blue, humpback whales)
      Family Balaenidae (right whales)
      Family Neobalaenidae (pygmy right whale)
      Family Eschrichtiidae (gray whale)
    Suborder Odontoceti (toothed whales)
      Family Delphinidae (oceanic dolphins)
      Family Phocoenidae (porpoises)
      Family Monodontidae (beluga and narwhal)
      Family Platanistidae (river dolphins)
      Family Ziphiidae (beaked whales)
      Family Physeteridae (sperm whales)
  Order Sirenia (sea cows, manatees, and dugongs)
  Order Proboscidea (elephants)
  Order Perissodactyla (horses, tapirs, and rhinoceroses)
  Order Hyracoidea (hyraxes)
  Order Tubulidentata (aardvark)
  Order Artiodactyla (pigs, sheep, giraffes, camels, etc.)
  Order Pholidota (pangolins)
  Order Rodentia (mice, rats, squirrels, porcupines, etc.)
  Order Lagomorpha (rabbits, pikas)
  Order Macroscelidea (elephant shrews)

SPECIFIC CLASSIFICATION OF ORDER CETACEA

The classification of whales is still somewhat fluid. Problems remain in the more diversified genera. The species of *Balaenoptera* are in flux, particularly the small species (*acutorostrata*, *bonaerensis*, *borealis*, *brydei*, and *edeni*). In the oceanic dolphins, the genera *Delphinus*, *Sousa*, and *Tursiops* are also in flux. The problematic species that have been grouped here into the river dolphins (family Platanistidae) are sometimes split by others into four separate families of uncertain relationships. The number of species is shown in square brackets.

ORDER **CETACEA** (whales and porpoises) [86]
Suborder **Mysticeti** (baleen whales) [14]
   Family **Balaenidae** [4]
     Balaena mysticetus   bowhead, Greenland right whale (archaic)
     Eubalaena glacialis   North Atlantic right whale
     Eubalaena japonica   North Pacific right whale
     Eubalaena australis   southern right whale
   Family **Neobalaenidae** [1]
     Caperea marginata   pygmy right whale
   Family **Eschrichtiidae** [1]
     Eschrichtius robustus   gray whale
   Family **Balaenopteridae** [8]
     Balaenoptera acutorostrata   minke whale
     Balaenoptera bonaerensis   Antarctic minke whale
     Balaenoptera borealis   sei whale
     Balaenoptera brydei   Bryde's whale
     Balaenoptera edeni   Eden's whale
     Balaenoptera musculus   blue whale
     Balaenoptera physalus   fin whale, finback, razorback
     Megaptera novaeangliae   humpback whale
Suborder **Odontoceti** (toothed whales, including porpoises) [72]
   Family **Physeteridae** [3]
     Physeter catodon   sperm whale
     Kogia breviceps   pygmy sperm whale
     Kogia sima   dwarf sperm whale
   Family **Monodontidae** [2]
     Delphinapterus leucas   beluga, white whale, belukha
     Monodon monoceros   narwhal
   Family **Ziphiidae** [20]
     Berardius arnuxii   Arnoux's beaked whale
     Berardius bairdii   Baird's beaked whale
     Hyperoodon ampullatus   northern bottlenose whale
     Hyperoodon planifrons   southern bottlenose whale
     Mesoplodon bidens   Sowerby's beaked whale
     Mesoplodon bahamondi   Bahamonde's beaked whale
     Mesoplodon bowdoini   Andrews' beaked whale
     Mesoplodon carlhubbsi   Hubbs's beaked whale
     Mesoplodon densirostris   Blainville's beaked whale
     Mesoplodon europaeus   Gervais's beaked whale
     Mesoplodon ginkgodens   ginkgo-toothed beaked whale
     Mesoplodon grayi   Gray's beaked whale
     Mesoplodon hectori   Hector's beaked whale

Mesoplodon layardii   strap-toothed whale
Mesoplodon mirus   True's beaked whale
Mesoplodon pacificus   Longman's beaked whale
Mesoplodon peruvianus   pygmy beaked whale
Mesoplodon stejnegeri   Stejneger's beaked whale
Tasmacetus shepherdi   Shepherd's beaked whale
Ziphius cavirostris   Cuvier's beaked whale

Family **Delphinidae** [36]
Cephalorhynchus commersonii   Commerson's dolphin
Cephalorhynchus eutropia   black dolphin
Cephalorhynchus heavisidii   Heaviside's dolphin
Cephalorhynchus hectori   Hector's dolphin, whitefront dolphin
Delphinus capensis   long-beaked common dolphin
Delphinus delphis   common dolphin, saddleback porpoise, white-bellied porpoise, short-beaked common dolphin
Delphinus tropicalis   Arabian common dolphin
Feresa attenuata   pygmy killer whale
Globicephala macrorhynchus   short-finned pilot whale, pothead, pilot whale
Globicephala melas   long-finned pilot whale, pothead, pilot whale, blackfish
Grampus griseus   Risso's dolphin, grampus
Lagenodelphis hosei   Fraser's dolphin, short-snouted whitebelly
Lagenorhynchus acutus   Atlantic white-sided dolphin
Lagenorhynchus albirostris   white-beaked dolphin
Lagenorhynchus australis   Peale's dolphin
Lagenorhynchus cruciger   hourglass dolphin
Lagenorhynchus obliquidens   Pacific white-sided dolphin
Lagenorhynchus obscurus   dusky dolphin, southern striped porpoise
Lissodelphis borealis   northern right whale dolphin
Lissodelphis peronii   southern right whale dolphin
Orcaella brevirostris   Irrawaddy River dolphin
Orcinus orca   killer whale
Peponocephala electra   melon-headed whale, electra
Pseudorca crassidens   false killer whale
Sotalia fluviatilis   tucuxi
Sousa chinensis   Indo-Pacific humpback dolphin, Pacific sousa
Sousa plumbea   Indian hump-backed dolphin, Indian sousa
Sousa teuszii   Atlantic hump-backed dolphin, West African sousa
Stenella attenuata   pantropical spotted dolphin
Stenella clymene   Clymene dolphin
Stenella coeruleoalba   striped dolphin, streaker
Stenella frontalis   Atlantic spotted dolphin
Stenella longirostris   spinner dolphin
Steno bredanensis   rough-toothed dolphin
Tursiops aduncus   Indian Ocean bottlenose dolphin
Tursiops truncatus   bottlenose dolphin

Family **Phocoenidae** [6]
Neophocaena phocaenoides   finless porpoise
Phocoena dioptrica   spectacled porpoise
Phocoena phocoena   harbor porpoise
Phocoena sinus   vaquita, Gulf of California harbor porpoise, cochito

Phocoena spinipinnis   Burmeister's porpoise

Phocoenoides dalli   Dall's porpoise

Family **Platanistidae** [5]

Inia geoffrensis   Amazon river dolphin, boutu, bouto

Lipotes vexillifer   Yangtze river dolphin, Chinese river dolphin, white flag porpoise,
pei c'hi, baiji

Platanista gangetica   Ganges river dolphin, Ganges susu

Platanista minor   Indus river dolphin, Indus susu

Pontoporia blainvillei   franciscana, La Plata river dolphin

APPENDIX 2.

CAREERS IN MARINE MAMMAL SCIENCE

WHAT IS MARINE MAMMAL SCIENCE?

There are about 100 species of aquatic or marine mammals that depend on freshwater or the ocean for part or all of their life. These species include pinnipeds, which are seals, sea lions, fur seals, and the walrus; cetaceans, which are baleen and toothed whales, ocean and river dolphins, and porpoises; sirenians, which are manatees, dugongs, and sea cows; and some carnivores, such as sea otters and polar bears. Marine mammal scientists try to understand these animals' genetic, systematic, and evolutionary relationships; population structure; community dynamics; anatomy and physiology; behavior and sensory abilities; parasites and diseases; geographic and microhabitat distribution; ecology; management; and conservation.

HOW DIFFICULT IS IT TO PURSUE A CAREER IN MARINE MAMMAL SCIENCE?

Competition for positions in marine mammal science is keen. However, working with marine mammals is appealing because of strong public interest in these animals and because the work is personally rewarding.

Marine mammal scientists are hired because of their skills as scientists, not because they like or want to work with marine mammals. A strong academic background in basic sciences, such as biology, chemistry, and physics, coupled with good training in mathematics and computers, is the best preparation for a career in marine mammal science. Persistence and diverse experiences make the most qualified individuals. Often developing a specialized scientific skill or technique, such as acoustics analysis, biostatistics, genetic analysis, or biomolecular analysis, provides a competitive edge.

No specific statistics are available on employment of students trained as marine mammal scientists. However, in 1990 the National Science Board reported some general statistics for employment of scientists within the United States: 75 percent of scientists with B.S. degrees were employed (43 percent of them held positions in science or engineering), 20 percent were in graduate school, and 5 percent were unemployed.

WHAT ARE TYPICAL SALARIES IN MARINE MAMMAL CAREERS?

Marine mammal scientists enter this field for the satisfaction of the work, not for the money-making potential of the career. Salaries vary greatly among marine mammal scientists, with government and industry jobs having the highest pay. Salary levels increase with years of experience and with graduate degrees, but generally they remain low considering the amount of experience and education needed. High competition in this field most likely will keep salaries at a modest level. A 1990 survey of 1,234 mammalogists conducted by the American Society of Mammalogists indicated that 42.7 percent of the respondents earned less than $40,000 per year. The salary range that included the most respondents (21.2 percent) was the $30,000 to $40,000 range.

WHAT TYPES OF JOBS INVOLVE MARINE MAMMALS?

Most jobs working with marine mammals are not as exciting or glamorous as popular television programs make them seem. Marine mammal studies often involve long, hard, soggy, sunburned days at sea, countless hours in a laboratory, extensive work on computers, hard labor such as hauling buckets of fish to feed animals, hours of cleanup, numerous reports, and tedious grant and permit applications.

As in other fields of science, jobs dealing with marine mammals vary widely. Examples of marine mammal jobs include researcher, field biologist, fishery vessel observer, laboratory technician, animal trainer, animal care specialist, veterinarian, whale-watch guide, naturalist, educator at any level, and government or private agency positions in legislative, management, conservation, and animal welfare issues. Many marine mammal scientists work with museum displays and collections, as curators, artists, illustrators, photographers, or filmmakers.

Answering the following questions will help you focus your interests and indicate which marine mammal scientists and facilities to contact for education, work experience, and job opportunities.

1. What specific areas are of interest to you (for example, anatomy, physiology, evolution, taxonomy, ecology, ethology, psychology, molecular biology, genetics, veterinary medicine, pathology, toxicology, biostatistics, management, conservation, museum curation, or education)?

2. What species or group of marine mammals is of interest to you (for example, cetaceans, sirenians, or marine carnivores)?

3. Do you desire a career involved in fieldwork or laboratory work?

4. Do you want a career involved with the care of animals, teaching, research, or legislative/policy matters?

5. Is working for government, industry, academia, oceanaria, museums, or private organizations or in self-employment best?

6. In what part of the world do you want to work?

For example, the manatee is an endangered species in Florida. It has a high mortality rate because of accidental entrapment in flood-control gates, collisions with speed boats, and loss of habitat. Local, state, and federal governments fund research on this species. Some local industries also are involved with the management of manatees. Therefore, people wanting to study manatees most likely should look for education and work experience at universities and research facilities in Florida.

WHO EMPLOYS MARINE MAMMAL SCIENTISTS?

A variety of international, federal, state, and local government agencies employ marine mammal scientists for positions in research, education, management, and legal/policy development. Federal agencies include the National Oceanic and Atmospheric Administration, the National Marine Fisheries Service, the Minerals Management Service, the U.S. Fish and Wildlife Service, the U.S. National Biological Service, the U.S. Navy, the Office of Naval Research, the U.S. Coast Guard, and the Marine Mammal Commission. Other federal agencies that work on marine-related issues include the U.S. National Park Service, the Army Corps of Engineers, the Environmental Protection Agency, the National Science Foundation, the National Aeronautics and Space Administration, the Department of State, and the Smithsonian Institution.

When oceanic operations, such as oil and gas exploration, production, and transportation, affect marine mammals, these industries often hire marine mammal experts. Because commercial fishing operations can conflict with marine mammal conservation, some fishing organizations hire marine mammal scientists. Many environmental, advocacy, and animal welfare organizations also hire marine mammal specialists. Oceanaria and zoos hire marine mammal specialists for veterinary care, husbandry, training, research, and education programs. Museums hire marine mammal specialists for educational programs, research, and curatorial positions.

WHAT EDUCATION IS NECESSARY TO BECOME A MARINE MAMMAL SCIENTIST?

A broad education is essential for finding employment in marine mammal science.

High School Studies   High school courses such as biology, chemistry, physics, mathematics, computer science, and language will provide a good educational base. Consult a guidance counselor for help in selecting course work. Good grades are essential for admission to a university.

Undergraduate Studies   Most entry-level marine mammal jobs require a B.S. degree, with a major in biology, chemistry, physics, geology, or psychology. A minor in any science, computer science, mathematics, statistics, or engineering also can be helpful. Good language and technical writing skills are essential. Many people are surprised by the amount of writing involved in marine mammal professions. Because marine mammals are found worldwide, foreign language training often is useful as well.

A student must first become a scientist before specializing in marine mammals. Generally, undergraduate students will concentrate on a basic science curriculum and rarely have an opportunity to take courses related to marine mammal science. Specialization in marine mammals generally comes later through practical work experience or while working toward an advanced degree. In other words, if your B.S. degree program does not include courses in marine sciences, do not become discouraged. Concentrate on finding practical experience and/or a master's degree program with emphasis in marine mammal science. Maintaining a high grade point average as an undergraduate is very important to gaining admission to graduate school.

Graduate Studies   The master's degree is usually the first opportunity college students have to specialize in marine mammal science. Care should be taken to select an adviser with experience in the subject and a reputable university with a diverse curriculum that will enable a focus on marine mammal science.

Students who have dual majors or interdisciplinary training sometimes have more employment opportunities. Because the field of marine mammal science is so diverse, students who train in specialized areas have practical tools that may help them gain employment. For example, a graduate degree in statistics can be very useful for entering the field of population assessment. A degree in electrical engineering can be particularly useful for bioacoustic research. A graduate degree in environmental law can be important for developing a career in government policy making or conservation.

WHAT ADDITIONAL CAREER OPPORTUNITIES WILL A GRADUATE DEGREE PROVIDE?

Generally, jobs requiring a B.S. (or B.A.) degree offer little opportunity for self-directed work. With this degree, potential positions include animal care specialist, animal trainer, field technician, laboratory technician, and consultant for industry, as well as positions in government at the entry level.

The M.S. (or M.A.) degree can facilitate individual work with marine mammals, for example, designing research projects, developing management plans, supervising field or laboratory studies, or heading programs in education, husbandry, or training.

The acquisition of a Ph.D. or D.V.M. (or both) provides even more career opportunities, including design and management of field and laboratory research programs, university faculty positions, coordination of government and industry programs, and management positions in oceanaria or museums.

Years of practical work experience sometimes can substitute for a graduate degree, but the time required to advance is typically longer for a person without an advanced degree.

HOW DOES SOMEONE FIND A UNIVERSITY PROGRAM IN MARINE MAMMAL SCIENCE?

Very few universities offer a marine mammal science curriculum. To select an undergraduate university, visit campuses and talk with professors and students about career interests. Most university libraries or counseling centers have university catalogs to identify schools with particular curricula. In addition, several publications list graduate programs by state and discipline, list marine mammal scientists by address, or summarize areas of research by marine mammal scientists.

An interest in a certain marine mammal species may influence the geographic location of the graduate university selected (see *What Types of Jobs Involve Marine Mammals?*). However, in most instances the best university is determined by selecting a graduate adviser specializing in a particular field.

Students should consider applying to several graduate schools. Application deadlines vary, but typically applications should be submitted in January for admission into a graduate program the following fall. Many universities require graduate school applicants to take the Graduate Record Examination (GRE) and include the test scores with their applications.

HOW DOES SOMEONE FIND AN ADVISER FOR GRADUATE STUDIES IN MARINE MAMMAL SCIENCE?

Selecting an adviser for a graduate degree is a very important decision. An adviser will become a mentor and a career-long colleague and will help establish a network of scientific colleagues. An adviser also helps to obtain funds to support graduate student research and helps make contacts for future employment.

First, identify marine mammal scientists who are doing current research in your area of interest, their university affiliation, whether they have funds to support graduate students, and if they are accepting new students. Keep in mind that many government and industry scientists also have adjunct appointments at universities and can serve as co-advisers.

There are two ways to find potential advisers:

1. Find the names of authors in current scientific journals, such as *Marine Mammal Science, Aquatic Mammals, Journal of Mammalogy, Canadian Journal of Zoology, Journal of Zoology, Behavioral Ecology and Sociobiology,* and *Fisheries Bulletin,* or in recently published books on marine mammals. Scientists who publish may be in situations where they can accept graduate students.

2. Attend specialized scientific conferences on marine mammals hosted by professional societies, such as the Society for Marine Mammalogy, the International Marine Animal Trainers' Association, the European Association for Aquatic Mammals, the European Cetacean Society, the American Cetacean Society, or the International Association for Aquatic Animal Medicine. Dates and locations of these meetings are published in the newsletter or journal of each society. At these meetings, make a personal contact with a potential adviser and express your interest in doing graduate work with him or her. Follow up any good lead with a telephone call, a letter, or a visit.

Because there is competition for advisers in the field of marine mammal science, an adviser will select students from a pool of applicants. Students should realize that, unlike the case in undergraduate study, graduate school faculty do *not* have to advise students just because they are enrolled at their university. Students sometimes enroll at a university because of a well-known professor and assume they will have the opportunity to work under him or her. *Before* entering a graduate program, contact the professor and establish his or her willingness to serve as an adviser. If necessary, discuss the possibilities of financial support and decide on a potential research project. Choose a thesis research topic carefully so that it is practical, scientifically sound, and potentially fundable. Seek advice from others on this, perhaps in the form of a draft research proposal. At many universities, the adviser needs to notify the graduate school to approve an application. Many prospective graduate students with good grades and experience are rejected because they do not have an adviser working from inside the university to facilitate their acceptance.

Many graduate schools will not accept students without financial support. Graduate assistantship funds for marine mammal studies are rare, and most graduate programs have a limited number of teaching assistantships. Students should be prepared to support themselves or find research funds on their own.

HOW DOES SOMEONE GAIN PRACTICAL WORK EXPERIENCE WITH MARINE MAMMALS?

As a high school or undergraduate student, you can gain practical experience by volunteering at federal, state, or local organizations that work with marine mammals. For example, volunteer as a laboratory assistant for a research project with marine mammals or volunteer for the marine mammal stranding network in the United States. Also, oceanaria, zoos, and museums often have large volunteer or docent programs. This volunteer experience provides practical skills, an employer reference, and a network of contacts in the field of marine mammal science, and, most important, it helps to determine whether this type of work is personally appealing. Because they already have observed a volunteer's work habits and commitment, organizations often hire from their pool of volunteers. Many oceanaria, zoos, museums, and government agencies have internships that provide practical experience.

Many careers in marine mammal science require experience in the marine environment. SCUBA certification, boat-handling experience, or sea time can be helpful in securing employment in the field of marine mammal science.

HOW DOES SOMEONE BECOME A MARINE MAMMAL TRAINER?

Most marine mammal trainers start by volunteering at an oceanarium or zoo. Often people work in other departments, such as operations, maintenance, or education, before transferring to a job in animal training. For the best advice about a career in marine mammal training, contact the International Marine Animal Trainers' Association.

HOW DOES SOMEONE BECOME A MARINE MAMMAL VETERINARIAN?

To become a marine mammal veterinarian, follow the basic curriculum and schooling of other veterinarians, but try to gain practical experience with marine mammals by volunteering at an oceanarium or zoo. A few veterinary schools are developing specialized course work in the area of exotic animal medicine, including marine mammals. For more information, contact the American Veterinary Medical Association and the International Association for Aquatic Animal Medicine.

HOW DOES SOMEONE FIND OUT ABOUT JOB OPENINGS WITH MARINE MAMMALS?

Often a good source for job announcements is the personnel department of a specific agency. Also, the journal *Science* and the *Chronicle of Higher Education* list academic positions at junior colleges, colleges, and universities.

Many jobs are not announced but rather are filled by volunteers at an organization, by a graduate student of a colleague, through an informal interview at a scientific conference, or from a recommendation by a colleague. In addition to what you know, who you know is very important in finding a marine mammal job. Keeping an active network of marine mammal colleagues is valuable. Attending scientific conferences is very useful for maintaining the network and identifying job opportunities. Electronic bulletin boards, such as MARMAM or WhaleNet, announce upcoming jobs. When looking for a job, make that fact known in these informal networks of marine mammal scientists.

Many job opportunities occur from being in the right place at the right time. Controlling the right time is difficult, but you can obtain the appropriate education, be in the right place, and wait for the right time. For example, chances of developing a career in designing educational exhibits on marine mammals are greatly enhanced if a candidate has an M.S. degree and volunteers in the exhibits department of an oceanarium.

APPENDIX 3.

SOURCES OF ADDITIONAL INFORMATION ON CETACEANS

American Cetacean Society
P.O. Box 4416
San Pedro, California 90731

The American Cetacean Society provides fact sheets, distributes action letters and newsletters featuring information on current issues involving marine mammals, and publishes a national quarterly, the *Whalewatcher: Journal of the American Cetacean Society.*

Center for Marine Conservation
1725 DeSales Street, N.W.
Washington, D.C. 20036

The Center for Marine Conservation sponsors the Whale Protection Fund, which concentrates its activities in the areas of education and public information. The center also publishes the quarterly *Marine Conservation News.*

Cetacean Society International
190 Stillwold Drive
Wethersfield, Connecticut 06109

The Cetacean Society International issues the newsletter *Connecticut Whale*, which includes current accounts of International Whaling Commission activities, local education projects, and the programs of active conservation organizations.

European Cetacean Society
c/o Dr. Peter Evans
Department of Zoology
University of Oxford
South Parks Road, GB-Oxford OX1 3PS
United Kingdom

The European Cetacean Society was formed in 1987 at a meeting of 80 cetologists from 10 European countries. The society brings together people from European countries studying cetaceans in the wild, allowing collaborative projects with international funding. A newsletter is produced three times a year for members, reporting current research in Europe, recent publications and abstracts, reports of working groups, conservation issues, legislation and regional agreements, local news, and cetacean news from around the world.

Hebridean Whale and Dolphin Trust
28 Main Street
Tobermory, Isle of Mull, Argyll PA75 6NU
Scotland

The Hebridean Whale and Dolphin Trust pioneered the study of the whales, dolphins, and porpoises found in the waters of the Hebrides. The research also provides the materials for the trust's work in environmental education, helping to raise public awareness around the United Kingdom and educate visitors to this outstanding area about the wildlife that exists right on their doorstep.

Hubbs Marine Center
1700 South Shores Road
San Diego, California 92109

The Hubbs Marine Center provides public information on various fields of marine mammal research.

National Audubon Society
950 Third Avenue
New York, New York 10022

Study materials for school and youth groups, posters, charts, and slides are offered at reasonable cost by the National Audubon Society. Local chapters are focal points for specific inquiries. *Audubon Magazine*, published by the society, features articles and color photographs on wildlife and reports on environmental problems.

National Wildlife Federation
1412 16th Street, N.W.
Washington, D.C. 20036

The National Wildlife Federation distributes numerous periodicals, educational materials, a catalog of nature-related materials, and free fact sheets on marine mammals. The publications list and catalog are available on request. An annual *Conservation Directory* is published and sold by the organization. It contains a list of organizations, agencies, and officials concerned with natural resource use and management, bibliographic materials on conservation, and sources for audio-visual aids.

Naval Oceans System Center
Code 5103 Seaside
San Diego, California 92152

information regarding research on dolphin sound production and other aspects of dolphin biology is provided by the Naval Oceans System Center.

New York Zoological Society
Zoological Park
Bronx, New York 10460

The New York Zoological Society operates the New York Zoo, the New York Aquarium, and the Osborn Laboratories of Marine Sciences. The society provides public information on marine mammals and promotes educational programs and zoological research. *Animal Kingdom* is the official publication.

Smithsonian Institution Press
P.O. Box 960
Herndon, Virginia 20172-0960
1-800-782-4612

The Smithsonian Institution Press publishes many books about marine biology, including *A Field Guide to the Whales, Porpoises, and Seals from Cape Cod to Newfoundland*, by Steven K. Katona, Valerie Rough, and David T. Richardson—$17.95 (paperback) (ISBN 1-56098-333-7); *Biology of Marine Mammals*, by John E. Reynolds III and Sentiel A. Rommel—$75 (hardbound) (ISBN 1-56098-375-2); and *Conservation and Management of Marine Mammals*, by John R. Twiss Jr. and Randall R. Reeves—$60 (hardbound) (ISBN 1-56098-778-2).

Whale Museum
P.O. Box 1154
Friday Harbor, Washington 98250

The Whale Museum publishes *Cetus*, the journal of the Moclips Cetological Society, six times a year. Articles feature the biology of various species, strandings, workshops and educational programs, and conservation efforts.

PERIODICALS

Aquatic Mammals is published as several issues composing one annual volume by the European Association for Aquatic Mammals. Papers on the biology, medical care, conservation, and related investigations of aquatic mammals, especially captive ones, are emphasized. The current editor is Dr. Jeanette A. Thomas, of the Laboratory of Sensory Biology, Western Illinois University Regional Center, 3561 60th Street, Moline, Illinois 61265 (Jeanette_Thomas@ccmail.wiu.edu).

The *Canadian Journal of Fisheries and Aquatic Sciences* (formerly *Journal of the Fisheries Research Board*), issued monthly by the Fisheries Research Board of Canada, is an excellent source of technical papers on marine mammals as well as fisheries management and ocean science.

Cetology was composed of technical papers on the biology of marine mammals, with an emphasis on cetaceans. Issues were numbered separately and were published irregularly. They were available only through subscription from Biological System, Inc., which is now defunct.

Hvalradets Skrifters, a Norwegian journal, publishes many technical papers on whales; all are in English. It is issued at irregular intervals by the Norske Videnskaps-Akademi, Oslo (Universitetsforlaget, Box 307, Blindern, Oslo 3, Norway).

The *International Marine Mammal Scientists Directory* has been superseded by the *Membership Directory of the Society for Marine Mammalogy*, which is published periodically as a supplement to *Marine Mammal Science*. The directory lists individuals and organizations engaged in marine mammal research.

Responsibility for setting quotas for member whaling nations resides in the International Whaling Commission, The Red House, Station Road, Histon, Cambridge CB4 4NP, United Kingdom. *The International Whaling Commission Annual Report* presents the findings of the annual meetings, catch limits for member nations, status of whale stocks, and scientific papers emphasizing distribution, occurrence, and behavior of various cetacean species. Numerous technical papers on the biology and behavior of specific cetaceans are presented at the Scientific Committee Meeting and were published as the *Reports of the Scientific Committee*. In 1999 the International Whaling Commission began publishing the *Journal of Cetacean Research and Management*, which replaced the *Reports*.

Journal of Mammalogy, a publication of the American Society of Mammalogists, issues an index to articles it has published. Listings are by subject, locality, vernacular name, scientific name, and author. Dues for the journal and requests for special publications should be addressed to the Journal of Mammalogy, Allen Marketing and Management, P.O. Box 1897, Lawrence, Kansas, 66044-8897.

The Marine Mammal Commission is an independent agency providing annual reports and copies of pertinent articles describing current research and the status of marine mammals. A list of marine mammal contract reports is available directly from the commission, at 1825 Connecticut Avenue, N.W., Suite 512, Washington, D.C. 20009.

Marine Mammal Science, a quarterly journal published by the Society for Marine Mammalogy, an international organization of marine mammalogists, is devoted to authoritative, technical papers on all aspects of marine mammal biology and includes book reviews and letters with scientific commentary on issues affecting marine mammals or marine mammal science. For subscription information, write to Dr. K. J. Frost, Alaska Department of Fish and Game, 1300 College Road, Fairbanks, Alaska 99701.

Bibliographies and information on whales, dolphins, porpoises, seals, and sea lions are available from the National Marine Fisheries Service, Marine Mammals and Endangered Species Division, Washington, D.C. 20235. This government agency cooperates with the Fish and Wildlife Service of the Department of the Interior in the management of marine mammals. Copies of the Endangered Species Act of 1973 and the Marine Mammal Protection Act of 1972 are available upon request. Published yearly as updates and every three years in comprehensive form, the *Marine Mammal Protection Act of 1972 Annual Report* describes the current activities of the National Marine Fisheries Service in regard to marine mammals and generally includes new information on marine mammal stocks. Original research reports and technical notes on investigations in fishery sciences are published in the quarterly *Fishery Bulletin*, issued by the National Oceanographic and Atmospheric Association branch of the service in Seattle, Washington. The bulletin includes some papers on marine mammals, such as those concerning the tuna-porpoise problem.

The *Newsletter of the Cetacean Specialist Group* is published irregularly (about once a year) by the International Union for Conservation of Nature and Natural Resources. It reports on endangered populations, new research projects, meetings, and monitoring results, with an emphasis on small cetaceans.

Oceanic Abstracts, a bimonthly publication, abstracts and indexes worldwide technical literature on the oceans, including marine biology, living and nonliving resources. It includes a year-end cumulative index.

Scientific Reports of the Whales Research Institute (1-3 Fukagawa Etchuzima, Koto-ku, Tokyo, Japan) was a collection that included technical papers emphasizing cetacean biology, but its publication has ceased. However, it has been replaced by *Scientific Reports of Cetacean Research*, which was to be published annually (since 1990; in English) by the Institute of Cetacean Research, Tokyo, Japan, but so far only one issue (1990) has come out.

SEAN Bulletin (Scientific Event Alert Network), published monthly by the Smithsonian Institution, U.S. National Museum of Natural History, Washington, D.C. 20560, reported strandings of cetaceans, pinnipeds, and sirenians worldwide as well as unusual biological occurrences up to 1982. (The biological section, including cetaceans, has been discontinued;

past issues are available from the National Technical Information Service, Springfield, Virginia 22161, stock no. PB81-9157.)

WHALE BIBLIOGRAPHIES

Allen, J. A. 1882. Preliminary lists of works and papers relating to the mammalian orders Cetacea and Sirenia. *Bulletin of the U.S. Geological and Geographical Survey* 6 (3): 399–562. Reprinted, New York: Arno Press, 1974. (1,013 references up to 1840.)

Bekierz, F. W. 1986. *Bibliographie über Wal—Bibliographien: Cetacea* (Bibliography on whales—bibliographies: Cetaceans). Courier Forschungsinstitut Senckenberg, vol. 88. Frankfurt am Main: Senckenbergischen Naturforschenden Gesellschaft.

Bird, J. E. 1977. *Whales, Whaling, Dolphins and Porpoises: An Annotated Bibliography.* Publication No. ACSSD-1. American Cetacean Society, San Diego, Calif.

———. 1983. An annotated bibliography of the published literature on the humpback whale (*Megaptera novaeangliae*) and the right whale (*Eubalaena glacialis/australis*), 1864–1980, part 5. Pp. 467–628 in Roger Payne, ed., *Communication and Behavior of Whales.* Boulder, Colo.: Westview Press.

Braham, H. W. 1988. An annotated bibliography of right whales, *Eubalaena glacialis*, in the North Pacific. Pp. 65–77 in R. L. Brownell, P. B. Best, and J. H. Prescott, eds., *Right Whales: Past and Present Status. Reports of the International Whaling Commission*, Special Issue 10. National Marine Mammal Laboratory, Northwest and Alaska Fisheries Center, NMFS, NOAA, Seattle, Wash.

Fodor, B. 1971. *The Sperm Whale* (Physeter catodon, *L.*): *A Bibliography.* Bibliography Series No. 25, Office of Library Services, U.S. Department of the Interior, Washington, D.C. (735 references on literature published from 1940 to 1970; available from the National Technical Information Service, Springfield, Va. 22161, stock no. PB-200-212.)

Forster, H., comp. 1985. *The South Sea Whaler.* An annotated bibliography of published historical, literary, and art material relating to whaling in the Pacific Ocean in the nineteenth century. Sharon, Mass.: Kendall Whaling Museum, and Fairhaven, Mass.: Edward J. Lefkowicz.

Gold, J. P. 1981. *Marine Mammals: A Selected Bibliography.* Revised 1979 edition. Marine Mammal Commission, 1625 I St., N.W., Washington, D.C. (647 titles; includes legal references, sources of information, and list of Marine Mammal Commission Reports; available from the National Technical Information Service, Springfield, Va. 22161, stock no. PB-82-104282.)

Heizer, R. F. 1968. A bibliography of aboriginal whaling. *Journal of the Society for the Bibliography of Natural History* 4, part 7 (January): 344–362. (233 titles.)

Holbrook, J. R. 1980. *Dolphin Mortality Related to the Yellowfin Tuna Purse Seine Fishery in the Eastern Tropical Pacific.* An annotated bibliography. Technical Bulletin No. 2, Porpoise Rescue Foundation. (457 entries; includes popular section.)

Jenkins, J. 1948. Bibliography of whaling. *Journal of the Society for the Bibliography of Natural History* 2, part 4 (3 November): 71–166. (Approximately 2,000 entries.)

Johnson, S. W. 1979. *An Annotated Catalog of Published and Unpublished Sources of Data on Populations, Life History, and Ecology of Coastal Marine Mammals of California.* Southwest Fisheries Center Administrative Report LJ-79-1. National Oceanic and Atmospheric Administration. National Marine Fisheries Service, Southwest Fisheries Center, La Jolla, Calif. 92038. (873 entries on distribution, population reproduction, mortality, and fishery interactions of pinnipeds and cetaceans.)

Oliver, Guy W., comp. 1987. *Bowhead Whale*, Balaena mysticetus, *Bibliography.* Outer Continental Shelf Study Report MMS 86-0059. (Copies available from the National Technical

Information Service, 5285 Port Royal Rd., Springfield, Va. 22161, stock no. PB-88-120688.)

———. 1986. *Gray Whale*, Eschrichtius robustus, *Bibliography*. Prepared for U.S. Department of the Interior, Minerals Management Service, Alaska OCS Region, by the Coastal Energy Research Laboratory, University of Maryland, Eastern Shore. (Categorizes research literature on gray whales according to major research topics; issues of concern to outer continental shelf oil and gas development; and geographic areas. Extensive cross-index. Copies and poster available from the National Technical Information Service, 5285 Port Royal Rd., Springfield, Va. 22161, stock no. PB-88-130166.)

Magnolia, L. R. 1977. *Whales, Whaling and Whale Research: A Selected Bibliography*. Whaling Museum, Cold Spring Harbor, N.Y. (l,000 titles; available from the Whaling Museum, P.O. Box 25, Cold Spring Harbor, N.Y. 11724.)

———. 1979. *Selected Bibliography of Whales and Whaling*. (79 titles for nonprofessionals; available from the Whaling Museum, P.O. Box 25, Cold Spring Harbor, N.Y. 11724.)

Mitchell, E. D., R. R. Reeves, and A. Evely. 1986. *Bibliography of Whale Killing Techniques*. Special Issue 7. Reports of the International Whaling Commission, The Red House, Station Road, Histon, Cambridge, United Kingdom.

Setzler-Hamilton, E. M., and Guy W. Oliver. 1991. *Right Whale*, Balaena glacialis, *a Bibliography*. National Technical Information Service. (5285 Port Royal Rd., Springfield, Va. 22161, stock no. PB-91-121236.)

Severinghaus, N. C. 1979. *Selected Annotated References on Marine Mammals of Alaska*. Northwest and Alaska Fisheries Center, National Marine Fisheries Service, 7600 Sand Point Way, N.E., Seattle, Wash. 98115. (564 citations of current work and prior studies.)

Sherbrooke W. C. 1978. *Jojoba: An Annotated Bibliographic Update*. University of Arizona, Office of Arid Lands Studies, Tucson. (Supplement to Arid Lands Resources Information Paper No. 5.)

Sherman, S. C. 1986. *Whaling Logbooks and Journals 1613–1927: An Inventory of Manuscript Records in Public Collections*. Revised and edited by J. M. Downey and V. M. Adams, also H. Pasternack. New York: Garland Publishing.

Skaptason, P. A. 1971. *The Fin Whale* (Balaenoptera physalus, *L.*): *A Bibliography*. Bibliography Series No. 26, Office of Library Services, U.S. Department of the Interior. (Available from the National Technical Information Service, Springfield, Va. 22161, stock no. PB-200-293.)

Truitt, D. 1974. *Dolphins and Porpoises: A Comprehensive Annotated Bibliography of the Smaller Cetacea*. Detroit, Mich.: Gale Research Co. (3,549 titles.)

Whitfield, W. K., Jr. 1971. *An Annotated Bibliography of Dolphin and Porpoise Families Delphinidae and Platanistidae*. Special Scientific Report No. 26, Florida Department of Natural Resources. (1,185 titles.)

Wood, D. A. 1977. *A Bibliography on the World's Rare, Endangered and Recently Extinct Wildlife and Plants*. Environmental Series No. 3, Oklahoma State University, Stillwater, Okla.

Wood, F. G. 1980. *Annotated Bibliography of Publications from the U.S. Navy's Marine Mammal Program*. Update 1 July 1980. Code 5104, Naval Ocean Systems Center, San Diego, Calif. 92152. (Titles arranged under headings of sound, physiology, health care, breeding behavior, open sea release, tagging, and hydrodynamics.)

GLOSSARY

anosmic Lacking the sense of smell.

artiodactyl A member of Artiodactyla, the order of hoofed mammals that have an even number of toes on each foot; cloven-hoofed animals such as the cow, deer, and goat.

artisanal Referring to activities that are pursued by artisans, people who are trained in an old manual skill that others have replaced with modern techniques; usually used in conjunction with or as a synonym of *aboriginal*.

asymptotic length The length that the growth curve of a species approaches but never equals; the maximum length of a species as determined by the growth curve. An asymptote is a straight line associated with a curve that it approaches infinitely closely but never meets.

belly flopping Behavior where a whale or dolphin emerges from the water more or less vertically and the descends onto its belly with considerable splashing and noise.

biopsy The act of removing a small sample of tissue from an individual in order to subject it to testing. Biopsies are commonly done in humans to indicate whether an abnormal growth is cancerous. Wild, free-living cetaceans are biopsied by means of a biopsy dart. The tip of the dart penetrates the skin and pulls out a small sample of skin and underlying blubber. Most biopsies are done using poles, harpoons, or crossbow darts.

bow riding Behavior involving a cetacean's use of the bow wave of a boat or a larger animal as an assist to locomotion. Bow riding is like surfing on a moving wave.

bow wave A displacement wave that is produced in front of a body going through the water.

breaching Behavior where a whale or dolphin emerges from the water more or less vertically, then descends onto its back with considerable splashing and noise.

brit An obscure term, of English origin, that is used to refer to "whales' food." It appears to have come from English whalers who discovered "sea snails" (pteropods) in the water and stomach of bowhead whales. It does not appear to have been applied to euphausiid shrimp, which fin and blue whales eat. *Brit* is also used to refer to larvae and juveniles of several fishes.

callosity In general, the quality or state of being hardened or thickened. The term has been applied to the unique wartlike skin growths on the head of right whales. Right whale callosities are slightly lighter in color than the body of the whale, but they form a habitat for whale lice, which are white and thus make the callosities stand out.

case The connective tissue bag in the head of a sperm whale that contains spermaceti. It can be up to 6 meters long and contain up to 2 metric tons of spermaceti.

cement See **teeth.**

cetologist A scientist who specializes in the study of whales.

cladistic [noun: cladistics] Referring to a method of classification employing phylogenetic hypotheses as its basis and using recentness of common ancestry alone as the criterion for grouping taxa.

common name See **vernacular name.**

confidence intervals Statistical range over which a statistical estimate is likely to occur because of the vagaries of sampling. Commonly given as the 95 percent confidence interval, which is the range over which 95 percent of the true values of the parameter can occur and still generate the same estimate.

crittercam A small video camera that can be attached to an animal and record what the animal does.

cookie-cutter shark See *Isistius brasiliensis.*

crown See **teeth.**

cyamids Members of Cyamidae, a family of amphipod crustaceans that occur on the skin of whales; also known as whale lice.

delayed implantation Occurs when a female ovulates and conception takes place (the egg is fertilized), but development is halted and the embryo does not implant in the uterus. This is one method of prolonging the gestation period to allow it to synchronize with the climatic seasons.

dentine See **teeth.**

dorsal mantle length See **mantle length.**

echolocation The act of an individual listening for echoes of sounds it has produced and using them as information about its environment. When it is done by humans, it is usually referred to as sonar (*sound navigation ranging*).

enamel See **teeth.**

euphausiid Member of the crustacean family Euphausiidae (or the suborder Euphausiacea), a group of shrimplike planktonic organisms.

fabulous Referring to a being of fables having no documented proof of existence.

flensing The act of stripping the blubber off a whale carcass. The whale then undergoes lemming. Compare **lemming.**

gam Herd or school of whales. This term appears to have been traditionally used to refer to a grouping of up to 20 sperm whales in particular but was applied, at various times, to all whales. This grouping has also been known as a pod.

GPS (acronym for global positioning system) GPS depends upon the receipt of a series of satellite signals. Knowing the precise position of the satellites, the receiver can then compute its position both laterally and vertically. That is to say, one can determine not only precise latitude and longitude from a GPS receiver but elevation as well. The GPS system has obvious advantages over the previous passive systems such as LORAN (*long range navigation*) A and C, which were used, respectively, during World War II and the 1970s. GPS is being rapidly utilized by shipboard, airborne, and land-based observers.

growth-layer group See **teeth.**

herd A large number of dolphins or whales feeding or migrating together. Used by Yankee sperm whalers to refer to a main body of whales comprising 50 to 100 individuals.

heterodont An animal having a differentiated dentition; that is, the teeth are divided into incisors, canines, premolars, and molars. Compare **homodont.**

homodont An animal having an undifferentiated dentition; that is, all the teeth are the same shape. Compare **heterodont.**

homologous Having a similar function and the same evolutionary origin.

Isistius brasiliensis A small shark known commonly as the cookie-cutter shark from the smooth circular wounds that it produces in its prey. The cookie-cutter shark swims toward a whale, bites into the skin, and then is turned around by the forward motion of the whale. This circular motion produces a plug of blubber like the plug of ice cream that is removed by an ice cream scoop.

These wounds were known to Antarctic whalers, but the animal that produces them was not. They referred to the animal as the demon whale-biter or DWB.

ivory Strictly speaking, the hard substance, not unlike bone, of which the teeth of most mammals chiefly consist; dentine. Ivory is commonly applied to any hard, bonelike material, including bone itself and substances of plant origin (vegetable ivory).

junk See **melon.**

keel See **postanal keel.**

knot A unit of speed equal to 1 nautical mile (1.85 kilometers) per hour. See **nautical mile.**

krill Shrimplike plankton, usually applied to the crustacean family Euphausiidae; krill forms the main portion of the food of baleen whales in some parts of the world.

leaping Behavior when a whale or a dolphin leaps into the air and enters the water cleanly (without undue splashing).

lemming Removal of the meat and viscera from a whale after it is flensed. Compare **flensing.**

life history Significant features of the life cycle through which an organism passes, with particular reference to strategies influencing survival and reproduction.

lobtailing Behavior of a whale or dolphin raising its tail in the air and forcefully hitting the surface of the water with it. This behavior is sometimes repeated many times. Lobtailing may be a form of communication or aggression.

longevity The average life span of the individuals of a population under a given set of circumstances.

mammals Vertebrates that have hair, a four-chambered heart, and a single bone in the lower jaw, nurse their young, and are warm-blooded. The term *mammal* is derived from the Latin *mamma,* "breast."

mantle length The standard measurement for the size of squid. It is measured along the dorsal surface of the mantle from the posterior tip of the body to the anterior edge of the mantle behind the head. Same as dorsal mantle length.

melon The term that whalers used to refer to the forehead of a pilot whale or dolphin. The melon is composed nearly entirely of clear fat and oil. When this body is excised from a pilot whale, it resembles a melon. In a sperm whale, the same body is called the junk.

melon oil The fine oil obtained by rendering the melon. It is a mixture of oil and wax esters and has a distinct aroma. That same oil also occurs in a fat body on the inside of the rear portion of the lower jaw. Melon and jaw oil were the economic bases of dolphin and pilot whale fisheries.

modern whaling Performed with steam- or diesel-powered vessels that mount a harpoon gun, usually on the bow. The harpoon gun fires a harpoon that is attached to a line and usually has an explosive head, which goes off inside the whale.

mysticete A member of Mysticeti, one of two suborders of whales, commonly known as the baleen whales. Recent studies have shown that some fossil mysticetes had teeth. Living mysticetes have teeth as fetuses but resorb them before birth.

nautical mile The distance equal to 1 minute of latitude; 6,080 feet; 1.51 statute miles; 1.85 kilometers.

neonatal Newborn; neonatal individuals have been born following a normal-length gestation period and have characteristics that classify them as at the beginning of the perinatal period of development. One such characteristic is flaccid (folded over or bent) dorsal fins.

octopus One of the molluscan class Cephalopoda (*cephalos* = head + *pes* = feet; squids and kin) distinguished by having eight arms (*octo* = eight + *pes* = feet). Octopodes do not usually have a distinct mantle or fins, as squids do.

odontocete A member of Ondontoceti, one of two living suborders of whales and dolphins, commonly known as the toothed whales. A few species lack erupted teeth as adults.

oocytes The cells in the female sex organ that turn into eggs.

open-boat whaling Whaling in which the whale was struck (harpooned) from an open boat, usually propelled by sail or oars. Open-boat whaling persisted into the twentieth century in a subsistence form.

pelagic Living in the open sea; oceanic; not associated with the shore or land; noncoastal.

Penella The generic name of a highly modified, parasitic, caligoid copepod crustacean that lives on cetaceans with its body buried into the blubber and its gills and reproductive organs hanging freely in the water column like a stalk of wheat.

perissodactyl A member of Perissodactyla, the order of hoofed ungulates that have an odd number of toes on each foot, including the horse, tapir, and rhinoceros.

plankton Literally, "that which is drifting"; organisms that drift with the ocean currents as opposed to nekton, animals that can swim about, or benthos, those that live on the bottom.

pod See **gam.**

porpoising A behavior of cetaceans when they swim rapidly and launch their bodies out of the water in a low-angled leap to breathe. In this way they do not interrupt their swimming pattern with pauses to come to the surface and breathe. This behavior can be extended to result in a spectacular series of leaps.

postanal keel The sexually dimorphic character that is developed in males of some species. It is a hypertrophy of the blubber of the ventral part of the caudal peduncle (tailstock) that results in the peduncle becoming deeper, like the keel of a sailboat.

pteropod Literally, "winged foot"; a planktonic group of mollusks also known as sea snails or sea butterflies. See **brit.**

rendering The process of boiling or steaming animal tissue in order to separate the fat and oil from the muscle, bone, and sinew. The places where this is accomplished are called rendering works.

root See **teeth.**

rorqual Baleen whales of the genus *Baleanoptera,* which includes the blue, fin, sei, Bryde's, and minke whales. These whales are equipped with distinctive abundant grooves that begin on the lower (ventral) surface of their throat and continue onto their chest. The only other member of the family Balaenopteridae and the only other whale with numerous ventral grooves, but not normally included as a rorqual, is the humpback.

rostrum A Latin word meaning "the beak or bill of a bird." Its usage has been expanded to apply to any beaklike structure, such as the elongated nose or muzzle of a whale. In cetology, the term *rostrum* is applied particularly to beaked whales.

rudimentary Referring to a structure that is just beginning to evolve. Compare **vestigial.**

school A term used by Yankee whalers to refer to a group of sperm whales numbering between 20 and 50; also known as a shoal.

sea serpent Any of a number of fabulous animals, usually in the shape of a snake or eel.

shark A nonscaly fish with multiple gill slits; one of the selachians. Sharks have a cartilaginous skeleton that sometimes becomes calcified.

sound, sounding Diving suddenly as if going to the bottom. This term was used in allusion to sailors measuring the depth of the water.

spermaceti Literally, "the sperm (semen) of a whale." This misnomer was applied to the oil that is found in the head of the sperm whale. It is a mixture of oil and wax that is liquid at body temperatures but solidifies in the cool air. The spermaceti is cooled and the oil pressed out of it, leaving a clear wax. This wax has a melting point of 44°C and was extremely popular for fine candles. The standard unit of illumination, one candlepower, is defined as the light produced by a candle composed of pure spermaceti.

sperm oil The oil obtained from the blubber and spermaceti organ of sperm whales.

spy hopping Behavior of a whale or dolphin sticking its head out of the water, pausing, and looking around. The term was first used by Yankee whalers in reference to this activity.

surf riding Dolphin behavior very similar to bow riding, where the dolphins use surf instead of bow waves as a source of energy.

taxon [pl., taxa] Any unit among a group of related organisms considered to be sufficiently distinct to be treated as a separate unit..

taxonomy The theory and practice of describing, naming, and classifying organisms.

teeth Appendages borne on the jaws and consisting of the crown, which is the part of the tooth that is erupted and sits above the gum line, and the root, which anchors the tooth to the socket, or alveolus. The teeth of young animals are hollow and filled with a mixture of blood vessel and nerves, which are known as the pulp. As the animal grows older, the pulp cavity is filled. The thin outer layer of the crown is made of an extremely hard substance called enamel. The outer layer of root consists of a bony material termed cement. The inner substance of both the crown and the root is termed dentine. Dentine and cement occur in layers. Combinations of these layers, which represent a year's growth, have been termed growth-layer groups and are used to estimate the age of cetaceans.

telescoping One of the ways in which a cetacean skull has been modified in adapting to the aquatic environment. The back part of the skull moves forward and the forward part moves backward. This has been likened to the way a telescope retracts or closes.

thigmotactic Referring to animals that enjoy close physical contact with other animals. Cats are particularly thigmotactic.

tusk A long tooth that projects from the mouth when the jaws are closed. The tusk is usually used for feeding or defense. The hippopotamus is one of the few mammals that have tusks that are completely hidden by the lips when the mouth is closed.

vernacular name The name that is used in the common language (vernacular) for an item, in our case, the name of a cetacean. The vernacular name is used as an alternative to the scientific, or Latin, name of a species. The vernacular name for *Balaena mysticetus* is "bowhead whale." In this case the vernacular name is actually in common use by people who routinely encounter the bowhead; thus it is a "common name" in the sense of its widespread use. In many cases, species that are rarely encountered by humans and people have not been given a common name. In those cases biologists have felt that the general public would feel uneasy with the scientific name so have coined a vernacular name, such as "True's beaked whale" for *Mesoplodon mirus*. Such a vernacular name cannot really be called a common name though.

vertebrates Animals with a backbone; usually divided into fish, amphibians, reptiles, birds, and mammals.

vestigial Referring to a structure that was once better developed in an ancestral species. Compare **rudimentary.**

whale oil Any oil that is obtained from a cetacean, including train oil, sperm oil, and melon oil.

REFERENCES

Amasaki, H., H. Ishikawa, and M. Daigo. 1989. Developmental changes of the fore- and hind-limbs in the fetuses of the southern minke whale, *Balaenoptera acutorostrata*. *Anatomischer Anzeiger* (Jena, Germany) 169: 145–148.

Andrews, R. C. 1921. *A remarkable case of external hind limbs in a humpback whale*. Occasional publication of the American Museum Novitates, no. 9.

Ash, C. E. 1962. *Whaler's Eye*. New York: Macmillan Co.

Beland, P., S. De Guise, and R. Plante. 1992. *Toxicology and Pathology of St. Lawrence Marine Mammals*. Final report, World Wildlife Fund Research Grant.

Bel'kovich, V. M., and A. V. Yablokov. 1963. Marine animals "share experience" with designers. *Nauka Zhizn* 30 (5): 61–64.

Berzin, A. A. 1972. *The Sperm Whale*. Israel Program for Scientific Translations. ISPT Cat. No. 60070 7, U.S. Department of Commerce, National Technical Information Service, TT 71-50152. (Translated from the original Russian, *Kashalot*, Tikhookeanskii Nauchno-Issledovatel'skii Institut Rybnovo Khozyaistva i Okeanografii, Moskva, 1971.)

Boschma, H. 1938. On the teeth and some other particulars of the sperm whale. *Temminckia* 3: 151–278.

Boyd, I. L., C. Lockyer, and H. D. Marsh. 1999. Reproduction in marine mammals. Pp. 218–286 in J. D. Reynolds III and S. A. Rommel, eds., *Biology of Marine Mammals*. Washington, D.C.: Smithsonian Institution Press.

Bruce, F. F. 1979. *The International Bible Commentary with the New International Version*. New York: Guideposts.

Bruemmer, F. 1989. Arctic treasures. *Natural History* 6 (89): 38–47.

Chittleborough, R. G. 1965. Dynamics of two populations of the humpback whale, *Megaptera movaeangliae* (Borowski). *Australian Journal of Marine and Freshwater Research* 16: 33–128.

Clapham, P. J., and J. G. Mead. 1999. *Megaptera novaeangliae*. Occasional publication of the American Society of Mammalogists, *Mammalian Species*, no. 604: 1–9.

Clarke, R. 1956. Sperm whales of the Azores. *Discovery Reports* 28: 237–298.

Collett, R. 1911–12. *Norges pattedyr*. Christiania, Norway: H. Aschehoug and Co. (W. Nygaard).

Cummings, W. C., and P. O. Thompson. 1971. Underwater sounds from the blue whale, *Balaenoptera musculus*. *Journal of the Acoustical Society of America* 50 (4): 1193–1198.

Davies, J. L. 1963. The antitropical factor in cetacean speciation. *Evolution* 17 (1): 107–116.

Dawson, W. W. 1980. The cetacean eye. Pp. 53–100 in L. M. Herman, ed., *Cetacean Behavior: Mechanisms and Functions*. New York: John Wiley and Sons.

Delage, Y. 1886. *Histoire du Balaenoptera musculus é choué suir la plage de Langrune*. Poitiers, France: Typographie Oudin.

Dral, A. D. G. 1972. Aquatic and aerial vision in the bottlenosed dolphin. *Netherlands Journal of Sea Research* 5: 510–513.

———. 1975. Vision in cetacea. *Journal of Zoo Animal Medicine* 6 (6): 17–21.

Ellis, R. 1991. *Men and Whales*. New York: Alfred A. Knopf.

———. 1994. *Monsters of the Sea*. New York: Alfred A. Knopf.

Eschricht, D. F. 1869. Ni Tavler til Oplysning af Hvaldyrenes bygning. *Kongelige Danske Vidensk-abernes Selskabs Skrifter*, 5te raekke (ser. 5), *Naturvidenskabelig og Mathematisk Afdeling 9*, bind 1: 1–14.

Eschricht, D. F., and J. Reinhardt. 1861. Om nordhvalen (*Balaena mysticetus* L.); navning med Hensyn til diens Udbredning i Fortiden og Nutiden of til dens ydre og indre Saekjender. Translated as On the Greenland right whale (*Balaena mysticetus*). Pp. 1–150 in W. H. Flower, ed., *Recent Memoirs on the Cetatcea by Professors Eschricht, Reinhart and Lilljeborg*. London: Ray Society.

Evans, P. G. H. 1976. An analysis of sightings of cetacea in British waters. *Mammal Review* 6 (1): 5–14.

Fitzsimons, F. W. 1919–20. *The Natural History of South Africa: Mammals*. London: Longmans, Green and Co.

George, J. C., J. Bada, J. Zeh, L. Scott, S. E. Brown, and T. O'Hara. 1998. *Preliminary Age Estimates of Bowhead Whales via Aspartic Acid Racemization*. International Whaling Commission, Report of the Scientific Committee, SC/50/AS10.

George, J. C., R. S. Suydam, L. M. Philo, T. F. Albert, J. E. Zeh, and G. M. Carroll. 1995. Report of the spring 1993 census of bowhead whales, *Balaena mysticetus*, off Point Barrow, Alaska, with observations on the 1993 subsistence hunt of bowhead whales by Alaska Eskimos. *International Whaling Commission, Report* 45: 371–384. (P. 380 shows jade harpoon points estimated at over 75 years of age, taken from a bowhead.)

Geraci, J. G., and V. J. Lounsbury. 1993. *Marine Mammals Ashore—A Field Guide to Strandings*. Texas A & M University Sea Grant Publication.

Geyh, M. A., and H. Schleicher. 1990. *Absolute Age Determination: Physical and Chemical Dating Methods and Their Application*. New York: Springer-Verlag.

Goodall, R. N. P., and A. R. Galeazzi. 1986. Recent sightings and strandings of southern right whales off subantarctic South America and the Antarctic Peninsula. Pp. 173–176 in R. L. Brownell, P. B. Best, and J. H. Prescott, eds., *Right Whales: Past and Present Status*. *International Whaling Commission, Reports*, Special Issue 10.

Gray, D. 1882. Notes on the characters and habits of the bottlenose whale (*Hyperoodon rostratus*). *Proceedings of the Zoological Society of London*, 726–731.

Gray, J. E. 1868. On the pelvis and hind limbs of Whales. *Annals of Natural History* 2: 79.

Grubs, D. 1997. Interpretation of killer whale/white shark encounter. Florida Museum of Natural History, Shark Research web page.

Guyton, A. G. 1964. *Textbook of Medical Physiology*. 2d ed. Philadelphia: W. B. Saunders Co.

Harmer, S. F. 1928. The history of whaling. *Linnean Society of London, Proceedings* 140: 51–95.

Harwood, J., M.-P. Heide-Jorgensen, E. Helle, M. Kingley, R. J. Law, M. Olsson, and J. Schwarz. 1990. Report of the Joint Meeting of the Working Group on Baltic Seals and the Study Group on the Effects of Contaminants on Marine Mammals. International Council for the Exploration of the Sea, Marine Mammals Committee, C.M. 1990/N:14.

Hentschel, E. 1912. *Die Meeressäugetiere*. Leipzig: Theod. Thomas, Verlag.

Hentschel, E. 1937. Naturgeschichte nordatlantischen Wale und Robben. Pp. 1–54 in H. Lübbert and E. Ehrenbaum, eds., *Handbuch der Seefischerei Nordeuropas*, vol. 3. Stuttgart.

Herman, L. M., M. F. Peacok, M. P. Yunker, and C. J. Madsen. 1975. Bottlenose dolphin: Double-slit pupil yields equivalent aerial and underwater diurnal acuity. *Science* 189: 650–652.

Hofman, R. J. 1990. Cetacean entanglement in fishing gear. *Mammal Review* 20 (1): 53–64.

Hooker, S. K., and R. W. Baird. 1999. Deep-diving behaviour of the northern bottlenose whale, *Hyperoodon ampullatus* (Cetacea: Ziphiidae). *Royal Society of London, Proceedings* 266: 671–676.

Hunter, J. 1787. Observations on the structure and economy of whales. *Philosophical Transactions of the Royal Society of London* 77: 371–450.

International Whaling Commission. 1994. Report of the Scientific Committee, Annex R, report of ad hoc working group on the publication of whaling statistics. *Report of the International Whaling Commission* 44: 200.

Jurasz, C. M., and V. P. Jurasz. 1979. Feeding modes of the humpback whale, *Megaptera novaeangliae*, in southeast Alaska. *Scientific Reports of the Whales Research Institute* 31: 69–83.

Kanwisher, J. K., and S. H. Ridgway. 1983. The physiological ecology of whales and porpoises. *Scientific American* 248 (6): 110–117, 119–120.

Kastelein, R. A., T. Dokter, and P. Zwart. 1990. The suckling of a bottlenose dolphin calf (*Tursiops truncatus*) by a foster mother, and information of transverse birth bands. *Aquatic Mammals* 16: 134–138.

Kasuya, T. 1998. Finless porpoise—*Neophocoena phocaenoides* (G. Cuvier, 1829). Pp. 411–442 in S. H. Ridgway and R. Harrison, eds., *Handbook of Marine Mammals*, vol. 6: *The Second Book of Dolphins and the Porpoises*. San Diego: Academic Press.

Kasuya, T., and R. L. Brownell. 1979. Age determination, reproduction, and growth of franciscana dolphin, *Pontoporia blainvillei*. *Scientific Reports of the Whales Research Institute* 31: 45–67.

Kasuya, T., and H. Marsh. 1984. Life history and reproductive biology of the short-finned pilot whale, *Globicephala macrorhynchus*, off the Pacific coast of Japan. Pp. 259–310 in W. R. Perrin, R. L. Brownell Jr., and D. P. DeMaster, eds., *Reproduction of Whales, Dolphins and Porpoises*. *Reports of the International Whaling Commission*, Special Issue 6.

Kawamura, A. 1980. A review of food of balaenopterid whales. *Scientific Reports of the Whales Research Institute* 32: 155–198.

Kellogg, W. N. 1961. *Porpoises and Sonar*. Chicago: University of Chicago Press.

Kellogg, W. N., and R. Kohler. 1952. Reactions of the porpoise to ultrasonic frequencies. *Science* 116 (3010): 250–252.

Krzyszkowska, Olga. 1990. *Ivory and Related Materials—an Illustrated Guide*. Classical Handbook 3, Bulletin Supplement 59, Institute of Classical Studies (31–34 Gordon Square, London, WC1H 0PY; SBN 900587 62 8, ISSN 0951-2586).

Kuznetzov, V. B. 1990. Chemical sense of dolphins: Quasi-olfaction. Pp. 481–503 in J. A. Thomas and R. A. Kastelein, eds., *Sensory Abilities of Cetaceans*. New York: Plenum Press.

Lang, T. G., and K. S. Norris. 1966. Swimming speed of a Pacific bottlenose porpoise. *Science* 151 (3710): 588–590. (Recorded top speed of 29.9 kph [16.1 kt] for 7.5 seconds and 21.9 kph [11.8 kt] for 50 seconds.)

Le Boeuf, N. R., ed. 1998. *Marine Mammal Protection Act of 1972, Annual Report, January 1, 1997, to December 31, 1997*. U.S. Department of Commerce, National Oceanic and Atmospheric Administration, National Marine Fisheries Service, Office of Protected Resources.

Lillie, D. G. 1915. Cetacea. British Antarctica ("Terra Nova") Expedition, Natural History Report. *Zoology* 1 (3): 85–124 (British Museum of Natural History).

Lockyer, C. 1976. Body weights of some species of large whales. *Journal du Conseil permanent international pour l'exploration de la mer* 36 (3): 259–273.

Lomax, Alan. 1960. *The Folk Songs of North America*. New York: Doubleday and Co.

Luoma, J. R. 1989. Doomed canaries of Tadoussac. *Audubon*, March, 92–97.

McCormick, J. G., E. G. Wever, S. H. Ridgway, and J. Palin. 1980. Sound reception in the porpoise as it relates to echolocation. Pp. 449–467 in R.-G. Busnel and J. F. Fish, eds., *Animal Sonar Systems*. NATO Advanced Study Institutes Series, Series A: Life Sciences, vol. 28. New York: Plenum Press.

MacDill, M. 1929. Game laws for world's largest beasts. *Science News-Letter* 16 (442): 187–189.

Marine Mammal Commission. 1998. *Marine Mammal Commission, Annual Report to Congress, 1997*. Marine Mammal Commission, Bethesda, Md.

Mead, J. G. 1986. Twentieth-century records of right whales (*Eubalaena glacialis*) in the northwestern Atlantic Ocean. Pp. 109–119 in R. L. Brownell, P. B. Best, and J. H. Prescott, eds., *Right Whales: Past and Present Status. International Whaling Commission, Reports*, Special Issue 10.

Mead, T. 1961. *Killers of Eden: The Story of the Killer Whales of Twofold Bay*. Sydney: Angus and Robertson.

Meyer, C. R. 1976. *Whaling and the Art of Scrimshaw*. New York: H. Z. Walck.

Milinkovitch, M. C., G. Orti, and A. Meyer. 1993. Revised phylogeny of whales suggested by mitochondrial ribosomal sequences. *Nature* 361: 346–348.

Mitchell, E. D. 1974. Trophic relationships and competition for food in northwest Atlantic whales. Pp. 123–133 in M. D. B. Burt, ed., *Proceedings of the Canadian Society of Zoologists Annual Meeting* (2–5 June 1974). Fredericton, New Brunswick: Canadian Society of Zoologists.

Mitchell, E. D., and A. N. Baker. 1980. Age of reputedly old killer whale, *Orcinus orca*, "Old Tom" from Eden, Twofold Bay, Australia. Pp. 143–154 in W. F. Perrin and A. C. Myrick Jr., eds., *Age Determination of Toothed Whales and Sirenians. International Whaling Commission, Reports*, Special Issue 3.

Montagu, G. 1821. Description of a species of *Delphinus*, which appears to be new. *Memoirs of the Wernerian Natural History Society* 3: 75–82.

Morris, R. J. 1986. The acoustic faculty of dolphins. Pp. 369–399 in M. M. Bryden and R. J. Harrison, eds., *Research on Dolphins*. New York: Oxford University Press.

Muizon, C. de, D. P. Domning, and M. Parrish. 1999. Dimorphic tusks and adaptive strategies in a new species of walrus- like dolphin (Odobenocetopsidae) from the Pliocene of Peru. Comptes Rendus Academie des Sciences, Paris. *Sciences de la terre et des planètes/Earth & Planetary Sciences* 329: 449–455.

Nemoto, T. 1963. New records of sperm whales with protruded rudimentary hind limbs. *Scientific Reports of the Whales Research Institute* 17: 79–81.

Nerini, M. 1984. A review of gray whale feeding ecology. Pp. 423–450 in M. L. Jones, S. L. Swartz, and J. S. Leatherwood, eds., *The Gray Whale—Eschrichtius robustus*. San Diego: Academic Press.

Norris, K. S. 1968. The evolution of acoustic mechanisms in odontocete cetaceans. Pp. 297–324 in E. T. Drake, ed., *Evolution and Environment*. New Haven, Conn.: Yale University Press.

Ogawa, T. 1953. On the presence and disappearance of the hind limb in the cetacean embryos. *Scientific Reports of the Whales Research Institute* 8: 127–132.

Ohlin, A. 1893. Some remarks on the bottlenose-whale. *Lunds Universitets Arsskrift* 29: 1–13.

Ohsumi, S. 1965. A dolphin (*Stenella caeruleoalba*[sic]) with protruded rudimentary hind limbs. *Scientific Reports of the Whales Research Institute* 19: 135–136.

O'Leary, M. A., and J. H. Geisler. 1999. The position of Cetacea within Mammalia: Phylogenetic analysis of morphologic data from extinct and extant taxa. *Systematic Biology* 48 (3): 455–490.

Olesiuk, P. F., M. A. Bigg, and G. M. Ellis. 1990. Life history and population dynamics of resident killer whales (*Orcinus orca*) in the coastal waters of British Columbia and Washington State. Pp. 209–243 in P. S. Hammond, S. A. Mizroch, and G. P. Donovan, eds., *Individual Recognition of Cetaceans: Use of Photo-identification and Other Techniques to Estimate Population Parameters*. *International Whaling Commission, Reports*, Special Issue 12.

Omura, H. 1978. Preliminary report on morphological study of pelvic bones of the minke whale from the Antarctic. *Scientific Reports of the Whales Research Institute* 30: 271–279. (Found femur in 25 percent of sample, 13 specimens.)

Omura, H., S. Ohsumi, T. Nemoto, K. Nasu, and T. Kasuya. 1969. Black right whales in the North Pacific. *Scientific Reports of the Whales Research Institute* 21: 1–78.

Payne, K., P. Tyack, and R. Payne. 1983. Progressive changes in the songs of humpback whales (*Megaptera novaeangliae*): A detailed analysis of two seasons in Hawaii. Pp. 9–57 in R. Payne, ed., *Communication and Behavior of Whales*. AAAS Selected Symposia Series, 76.

Pecchioni, P., and C. Peoldi. 1998. Sperm whale spotted attacking megamouth shark. Florida Museum of Natural History, Shark Research web page.

Penniman, T. K. 1952. Pictures of ivory and other animal teeth, bone and antler, with a brief commentary on their use in identification. *Pitt Rivers Museum* (Oxford University), *Occasional Papers in Technology* 5: 1–40.

Perkins, J., and H. Whitehead. 1983. The legend of the swordfish and the thresher. *Whalewatcher* 17 (1): 10–15.

Perryman, W. L., M. A. Donahue, J. L. Laake, and T. E. Martin. 1997. Gray whale day/night migration rates determined with thermal sensors. Unpublished paper presented at the 1997 IWC Scientific Committee Meeting, SC/49/AS12.

Philbrick, Nathaniel. 2000. *In the Heart of the Sea—the Tragedy of the Whaleship Essex*. New York: Viking.

Pitman, R. L., and S. J. Chivers. 1999. Terror in black and white. *Natural History* 107 (10): 26–29.

Pitman, R. L., A. Aguaya, J. Urban. 1987. Observations of an unidentified beaked whale (*Mesoplodon* sp.) in the eastern tropical Pacific. *Marine Mammal Science* 3 (4): 345–352.

Porsild, M. P. 1922. Scattered observations on narwhals. *Journal of Mammalogy* 3 (1): 8–13, pl. 1.

Pryor, K. 1990. Concluding remarks on vision, tactition, and chemoreception. Pp. 561–569 in J. Thomas and R. Kastelein, eds., *Sensory Abilities of Cetaceans*. New York: Plenum Press.

Reeves, R. R., and J. G. Mead. 1999. Marine mammals in captivity. Pp. 412–436 in J. R. Twiss and R. R. Reeves, eds., *Conservation Management of Marine Mammals*. Washington, D.C.: Smithsonian Institution Press.

Reyes, J. C., K. v. Waerebeek, J. C. Cardenas, and J. L. Yanez. 1995. *Mesoplodon bahamondei* sp. n. (Cetacea, Ziphiidae), a new living beaked whale from the Juan Fernandez Archipelago, Chile. *Boletin Museo Nacional de Historia Natural de Chile* 45: 31–44.

Rice, D. W. 1989. Sperm whale. Pp. 177–233 in S. H. Ridgway and R. Harrison, eds., *Handbook of Marine Mammals*, vol. 4: *River Dolphins and the Larger Toothed Whales*. San Diego: Academic Press.

Ridgway, S. H. 1972. Homeostasis in the aquatic environment. Pp. 590–747 in S. H. Ridgway, ed., *Mammals in the Sea, Biology and Medicine*. Springfield, Ill.: C. C. Thomas.

Ridgway, S. H., and R. J. Harrison. 1986. Diving dolphins. Pp. 33–58 in M. M. Bryden and R. J. Harrison, eds., *Research on Dolphins*. New York: Oxford University Press.

Ridgway, S. H., B. L. Scronce, and J. Kanwisher. 1969. Respiration and deep diving in the bottlenose porpoise. *Science* 166: 1651–1654.

Robertson, Dougal. 1973. *Survive the Savage Sea*. New York: Praeger.

Rugh, D. J., M. M. Muto, S. E. Moore, and D. P. DeMaster, eds. 1999. *Status Review of the Eastern North Pacific Stock of Gray Whales*. NOAA Technical Memorandum NMFS-AFSC-103.

Scammon, C. M. 1874. *The Marine Mammals of the Northwestern Coast of North America—Described and Illustrated—Together with an Account of the American Whale Fishery*. San Francisco: John H. Carmany and Co.

Schevill, W. E., and B. Lawrence. 1949. Underwater listening to the white porpoise (*Delphinapterus leucas*). *Science* 109: 143–144.

Schulte, H. von W. 1916. Anatomy of a fetus of *Balaenoptera borealis*. *Memoirs of the American Museum of Natural History*, n.s., 1 (6): 389–502.

Scoresby, W., Jr. 1820. *An Account of the Arctic Regions, with a History and Description of the Northern Whale-Fishery*. Edinburgh: Archibald, Constable, and Co.

Slijper, E. J. 1962. *Whales*. London: Hutchinson and Co.

Slooten, E., and S. M. Dawson. 1988. Studies on Hector's dolphin, *Cephalorhynchus hectori:* A progress report. Pp. 325–338 in R. L. Brownell and G. P. Donovan, eds., *The Biology of the Genus* Cephalorhynchus. *International Whaling Commission, Reports*, Special Issue 9.

Smith, T. D., J. Allen, P. J. Clapham, P. S. Hammond, S. Katona, F. Larsen, J. Lien, D. Mattila, P. J. Palsboll, J. Sigurjonsson, P. T Stevick, and N. Oien. 1999. An ocean-basin-wide mark-recapture study of the North Atlantic humpback whale (*Megaptera novaeangliae*). *Mammal Science* 15 (1): 1–32.

Spector, W. S. 1956. *Handbook of Biological Data*. Prepared under the direction of the Committee on the Handbook of Biological Data, Division of Biology and Agriculture, the National Academy of Sciences, the National Research Council. Philadelphia: W. B. Saunders Co.

Starbuck, A. 1878. I. *History of the American Whale Fishery from Its Earliest Inception to the Year 1876*. United States Commission of Fish and Fisheries, part 4, Report of the Commissioner for 1875–1876. II. Appendix to Report of the Commissioner, Appendix A, the Sea Fisheries.

Struthers, J. 1889. *Memoir on the Anatomy of the Humpback Whale*, Megaptera longimana. Edinburgh: Maclachlan and Stewart. Reprinted from the *Journal of Anatomy and Physiology*, 1887–89. (Gives nine pages to the hind limb and pelvic girdle.)

Tavolga, M. C., and F. S. Essapian. 1957. The behavior of the bottlenosed dolphin (*Tursiops truncatus*): Mating, pregnancy, parturition and mother-infant behavior. *Zoologica* 42: 11–31.

Townsend, C. H. 1914. The porpoise in captivity. *Zoologica* 1 (16): 289–299.

Tressler, D. K. 1923. *Marine Products of Commerce*. New York: Chemical Catalog Co.

Trident Society. 1930. *The Book of Navy Songs*. New York: Doubleday, Doran and Co.

Tyson, E. 1681. *Phocaena; or the Anatomy of the Porpus, Dissected at Gresham College*. London: B. Tooke.

Vangstein, E., ed. 1953. Fin whale with 6 foetuses. *Norske Hvalfangst-tidende* (Norwegian whaling gazette) 42 (12): 685–686.

———. 1964. One hundred years of Norwegian whaling. *Norsk Hvalfangst-tidende* (Norwegian whaling gazette) 53: 122–137.

Wallace, Richard L. 1994. *The Marine Mammal Commission Compendium of Selected Treaties, International Agreements, and Other Relevant Documents on Marine Resources, Wildlife, and the Environment*. 3 vols. Marine Mammal Commission. Washington, D.C.: U.S. Government Printing Office.

———. 1997. *The Marine Mammal Commission Compendium of Selected Treaties, International*

Agreements, and Other Relevant Documents on Marine Resources, Wildlife, and the Environment. First update, comp. Richard L. Wallace. Marine Mammal Commission, Washington, D.C.: U.S. Government Printing Office.

Waring, G. T., D. L. Palka, P. J. Clapham, S. Swartz, M. C. Rossman, T. V. N. Cole, L. J. Hansen, K. D. Bisack, K. D. Mullin, R. S. Wells, D. K. Odell, and N. B. Barros. 1999. *U.S. Atlantic and Gulf of Mexico Marine Mammal Stock Assessments—1999.* NOAA Technical Memorandum, NMFS-NE-153.

Watkins, W. A., M. A. Daher, K. M. Fristrup, and T. J. Howard. 1993. Sperm whales tagged with transponders and tracked underwater by sonar. *Marine Mammal Science* 9 (1): 55–67.

White, P. D., and S. W. Mathews. 1956. Hunting the heartbeat of a whale. *National Geographic*, vol. 110, July, 49–64.

Whitehead, H. 1985a. Why whales leap. *Scientific American* 252 (3): 84–88, 93.

———. 1985b. Humpback whale breaching. *Investigations on Cetacea* 17: 117–155.

Wieland, G. R. 1908. The conservation of the great marine vertebrate: Imminent destruction of the wealth of the seas. *Popular Science Monthly*, May, 425–430.

Wilson, B. 1995. The ecology of bottlenose dolphins in the Moray Firth, Scotland: A population at the northern extreme of the species' range. Ph.D. dissertation, University of Aberdeen, Scotland.

Wilson, E. A. 1907. *Mammalia (Whales and Seals), National Antarctic Expedition, 1901–1904, Natural History*, vol. 2: *Zoology (Vertebrata: Mollusca: Crustacea)*. London: Trustees of the British Museum.

Wursig, B., and C. Clark. 1993. Behavior. Pp. 157–199 in J. J. Burns, J. J. Montague, and C. J. Cowles. 1993. *The Bowhead Whale*. Society for Marine Mammalogy, Special Publication No. 2.

Yablokov, A. V., and L. S. Bogoslovskaya. 1984. A review of Russian research on the biology and commercial whaling of the gray whale. Pp. 465–485 in M. L. Jones, S. L. Swartz, and S. Leatherwood, eds., *The Gray Whale*—Eschrichtius robustus. San Diego: Academic Press.

Yablokov, A. V., and M. Olsson. 1989. *Influence of Human Activites on the Baltic Ecosystem*. Leningrad: Gidrometeoizdat.

TAXONOMIC INDEX

Page numbers in boldface indicate illustrations.

phocoenids (Phocoenidae). *See* porpoise

pilot whale (*Globicephala* spp.), 6; short-finned (*Globicephala macrorhynchus*), 6, **7**

platanistids (Platanistidae). *See* river dolphin

polar bear (*Ursus maritimus*), **126,** 127

porpoise (Phocoenidae), 3, 4, 7, 11

Protocetus, 95

puffing pig (Phocoenidae), 4

pygmy beaked whale (*Mesoplodon peruvianus*), 98

pygmy killer whale (*Feresa attenuata*), 7

pygmy right whale (*Caperea marginata*), 100–102

right whale (Balaenidae), **8, 9,** 22, **57,** 72, **76, 92,** 98–100, **99**

Risso's dolphin (*Grampus griseus*), 7

river dolphin (Platanistidae), 2, 3, 12, 140–141

Rodhocetus, 95

rorqual (*Balaenoptera*) 5, 58, 107–113

roundworm (*Stenurus*), 58

sea canaries (*Delphinapterus leucas*), 88

sea cows (Sirenia), 70

sea pig (Phocoenidae), 4

sea snails (pteropods), 72

sei whale (*Balaenoptera borealis*), 5, 72, 111–113

shark, 72

shark sucker (*Remora*), **57,** 58

shrimp, 70

sirenians (Sirenia), 2, 70

southern right whale (*Eubalaena australis*), 98

sperm whale (*Physeter catodon*), 10, 22, **48,** 70, **80,** 119–123, **120,** 143, 148, **149, 161, 162,** 164, 165

spinner dolphin (*Stenella longirostris*), **153**

squid, 70, 72

swordfish (*Xiphias*), 81

thresher sharks (*Alopias*), 81

toothed whale, 2, 95

tucuxi (*Sotalia fluviatilis*), 12

vaquita (*Phocoena sinus*), **153**

white shark (*Carcharodon carcharias*), 72

white whale. *See* beluga

zeuglodon. *See* archaeocetes

Zygorhiza, 96

SUBJECT INDEX

Page numbers in boldface indicate illustrations.

flippers, 25, 82
fluke catalog, **19**
flukes, 8–9, **9**, 25, 27, 45, 82
folds, vocal, 86
folktales, 160
follicle, 60
food, quantity of, 75
forty-barrel bulls, 79
fossil, 95
Foyn, Svend, 5, 145, 152
Free Willie (film), 164
fresh water, cetaceans living in, 12
fruit, dolphins eating, 70
funny bone, 26

gam, 79–80
Gardner, Erle Stanley, 164
genetics, 10
genetic sampling, 10
Gesner, Konrad, 21
gestation, 61, 65
glove finger, 9
grazing predator, 82
"Greenland's Icy Shores" (song), 163
"Greenland Whale Fishery" (song), 162
group: cohesiveness of, 91; composition of sperm whales in, 79

hair, 8, 23
hand, 26
Harderian glands, 58
harem school, 79
hearing, 41–43
heart, 32
heart beat, 51
heart rate, 51
Hedges, Mitchell, 109
herds, 79
heterodont, 95
hind flipper, 8
hind limb, 8, 26
hip, 26
Historia Animalium (Aristotle), 21
History of the American Whale Fishery (Starbuck), 143
homodont, 95
Hooker, Sascha, 132
humerus, 26
Hunter, John, 22
Hunter and the Whale, The (Van Der Post), 164
Hunting the Desert Whale (Gardner), 164
Huston, John, 164
hval, 4
hwael, 4

identification, 52–53
ilium, 26
individual recognition, 17, **18, 19**
infanticide, 83
intelligence, 59

International Agreement for the Regulation of Whaling, 155, 159
International Convention for the Regulation of Whaling, 155
International Whaling Commission, 159
International Whaling Convention, 159
ischium, 26
ivory, 148; boar, **149**; bone, 150–151; elephant, 148; elk, **149**; hippopotamus, **149**; imitation, 149–150; sperm whale, **149**; tagua nut, 148, **149**; types of, **149**; vegetable, **149**; walrus, 148–149, **149**

Jonah, 165

karyotype, 10
Kendall Whaling Museum, 165
Kennedy, John F., 165
keratin, 31
King Alfred of Wessex. *See* Alfred the Great of England

lacrimal gland, 58
lactation, 61, 67–68
language, 89
larynx, 33
Laura (earliest factory ship), 145
length, 22
life history, **38**
Linnaeus, 97
lobtailing, **69**
lone bull, 79
longevity, 36, 38, **39**
lung, 33; collapse of, 50

magnetic sense, 44
Magnus, Olaus, 21
mammal, 8
mammary glands, 7
mammary line, 8
mammary slit, 7
management, international, 157
manticore, 21
manus, 26
Marine Mammal Commission, 156
Marine Mammal Protection Act (1972), 156
mass strandings, 58, 91
mating, 61, 63
maturity, reproductive, 62
medication, 90
Melville, Herman, 82, 143
midwives, 67
Miencke (crewman on whaling ship), 5
Miencke's whales, 5
migration, 12, **13**; of beluga, 127; of California gray whale,

14; feeding on, 15; hazards of, 14; of humpback, 14
migratory routes, 13
milk, 7
Moby Dick (film), 164
Moby Dick (Melville), 82, 143, 163–164
Mocha Dick (nineteenth-century white sperm whale), 164
molecular genetics, 19
Monstro (character in *Pinocchio*), 164
mother-calf tie, 67
muscles, 29, **29**

nasal passage, 33
Naturalis Istoria (Pliny the Elder), 21
net fisheries for whales, 146
newborn whales and dolphins, 36
nitrogen, 49
noise pollution, 42
Norris, Kenneth S., 75
Norwegian Whaling Act, 154–155
nursery school, 79

oceanic circulation, 12
Old Dartmouth Historical Society, 165
olecranon process, 26
olfaction, 43–44
olfactory nerve, 44
Oligocene, 95
Ottar, 144
ovary, 60–61

pain, 44
Pakistan, 95
Panama Canal, 109
parasites, 56, 58
parentage, 10
passive acoustic techniques, 17
Peck, Gregory, 164
pelvic girdle, 26
Pequod (fictional whaling ship), 82, 163
Permians. *See* Biarmans
Peru, 98
Peruvian anchoveta fishery, 12–13
pesticide residues, 10
pests, 56
phalanges, 26
Philbrick, Nathaniel, 143
physical maturity, 37
pigmentation, of narwhals, **31**
Pinocchio (film), 164
Pitman, Robert, 98
plankton, 70
play, 77
Pliny the Elder, 21
Pliocene, 97
plugged tusk, of narwhals, 85
pod, 79–80